Comparative
Color Vision

ACADEMIC PRESS
SERIES IN COGNITION AND PERCEPTION

SERIES EDITORS:
Edward C. Carterette
Morton P. Friedman
Department of Psychology
University of California, Los Angeles

Stephen K. Reed: *Psychological Processes in Pattern Recognition*

Earl B. Hunt: *Artificial Intelligence*

James P. Egan: *Signal Detection Theory and ROC Analysis*

Martin F. Kaplan and Steven Schwartz (Eds.): *Human Judgment and Decision Processes*

Myron L. Braunstein: *Depth Perception Through Motion*

R. Plomp: *Aspects of Tone Sensation*

Martin F. Kaplan and Steven Schwartz (Eds.): *Human Judgment and Decision Processes in Applied Settings*

Bikkar S. Randhawa and William E. Coffman: *Visual Learning, Thinking, and Communication*

Robert B. Welch: *Perceptual Modification: Adapting to Altered Sensory Environments*

Lawrence E. Marks: *The Unity of the Senses: Interrelations among the Modalities*

Michele A. Wittig and Anne C. Petersen (Eds.): *Sex-Related Differences in Cognitive Functioning: Developmental Issues*

Douglas Vickers: *Decision Processes in Visual Perception*

Margaret A. Hagen (Ed.): *The Perception of Pictures, Vol. 1: Alberti's Window: The Projective Model of Pictorial Information, Vol. 2 Dürer's Devices: Beyond the Projective Model of Pictures*

Graham Davies, Hadyn Ellis and John Shepherd (Eds.): *Perceiving and Remembering Faces*

Hubert Dolezal: *Living in a World Transformed: Perceptual and Performatory Adaptation to Visual Distortion*

Gerald H. Jacobs: *Comparative Color Vision*

in preparation

John A. Swets and Ronald M. Pickett: *Evaluation of Diagnostic Systems: Methods from Signal Detection Theory*

Diana Deutsch (Ed.): *The Psychology of Music*

Comparative Color Vision

Gerald H. Jacobs

Department of Psychology
University of California, Santa Barbara

ACADEMIC PRESS

A SUBSIDIARY OF HARCOURT BRACE JOVANOVICH, PUBLISHERS
New York   London   Toronto   Sydney   San Francisco   1981

ACADEMIC PRESS, INC.
111 Fifth Avenue, New York, New York 10003

United Kingdom Edition published by
ACADEMIC PRESS, INC. (LONDON) LTD.
24/28 Oval Road, London NW1 7DX

Library of Congress Cataloging in Publication Data

Jacobs, Gerald H.
  Comparative color vision.

  (Cognition and perception series)
  Bibliography: p.
  Includes index.
  1. Color vision. 2. Physiology, Comparative.
I. Title. II. Series.
QP483.J3   599.01'823   81-17654
ISBN 0-12-378520-0   AACR2

PRINTED IN THE UNITED STATES OF AMERICA

81 82 83 84   9 8 7 6 5 4 3 2 1

Contents

Preface

A few years ago I received a telephone call from someone inquiring how he should go about studying color vision in the American elk (*Cervus canadensis*). Although the subject for study in this case was unusual, the request was not. As in other similar instances, I was unable to direct my caller to some convenient and up-to-date source that would provide what he needed—namely, information about the means by which color vision has been studied in nonhuman animals and about the outcomes of these studies for a variety of representative species. This monograph is the result of my efforts to provide such information.

It has been my experience that those who become interested in color vision in animals come from a variety of different educational backgrounds—from the traditional biological and behavioral sciences as well as from more applied fields. Accordingly, I thought it would be useful to include in this work sufficient tutorial information about color vision so that a relative newcomer would be able to make sense out of this area without having to search out still more background material. To provide this, basic information about the psychophysics of color vision and about the methods used to study color vision in animals is presented in Chapter 2, and coverage of the broad range of biological mechanisms responsible for color vision appears in the following chapter. Those already knowledgeable about these features of color vision will find the latter part of the book the most useful. Chapters 4 and 5 contain systematic reviews of studies of color vision in a wide selection of vertebrate species. In writing these reviews, I tried to interpret the data that were clear, note controversial issues, and provide suggestions where the record was unclear or missing. The final chapter is devoted to a discussion of two fascinating issues raised by studies of animal color vision: the evolutionary origins and the functional utility of color vision.

My laboratory has been involved in the study of animal color vision for a number of years. Consequently, in preparing this book I have drawn on our own work for illustrative purposes in a number of instances, often where the work of others might have sufficed. This was done for convenience only, not because our studies were necessarily more appropriate.

Many colleagues, students, and other friends, especially those in my own laboratory, have helped me to learn about animal color vision. I thank them all, but the following deserve specific mention for their direct contributions to this book: Don Anderson, Eileen Birch, Barbara Blakeslee, Steve Fisher, Jack Loomis, Mark McCourt, Ben Reese, and Max Snodderly. I also gratefully acknowledge the support for my research received from the United States Public Health Service through the National Eye Institute. I owe a special debt to Russell De Valois, who bears a central responsibility for my involvement with color vision in the first place, for demonstrating to me by word and example the utility of studying this topic from the joint vantages of behavior and electrophysiology. Of course, neither he nor any of the others named above should be held responsible for any of the residual curiosities of thought or expression in the pages that follow.

Chapter 1

Introduction

Color is such a ubiquitous and intrinsically fascinating aspect of man's visual world that it is hardly surprising that the scientific study of color vision has been vigorously pursued. As a result, a very substantial research literature has emerged. Although it is probably fair to conclude that considerable progress has been made toward understanding color vision, particularly with regard to its basic psychophysics and physiological mechanisms, it is clear that much remains to be thoroughly investigated. Aspects of color vision that warrant further investigation include the subject matter of this monograph—the extent, mechanisms, and nature of color vision among nonhuman species.

Before proceeding any further, it might be well to examine briefly the reasons for an interest in the phylogenetic aspects of color vision. Why would anyone dedicate many hundreds of hours to a study of color vision in, say, the goldfish? Although researchers only infrequently disclose the motivations of their work in formal reports, several underlying incentives for study of comparative color vision can be discerned. Perhaps the primary one derives from the strong belief that sensory systems are, in general, highly adaptive. Thus, according to this thinking, the features of a sensory system, its functional details, and its prominence as a processor of environmental information, are a direct reflection of the evolutionary pressures brought to bear on the ancestors of the species in question. For example, a species having an arboreal existence is likely to have achieved some degree of visual binocularity, presumably because binocular vision is a useful asset for a life that depends on accurate discriminations of depth and distance, whereas a closely related terrestrial species may very likely have more laterally placed eyes, perhaps thereby sacrificing substantial binocular overlap in order to achieve a more panoramic view of the world. It is precisely this kind of adaptiveness that some assume might explain the extensive phylogenetic variations in color vision, that might indicate why some species have received as a bequest of the evolutionary process excellent color vision while others have only the most miniscule color capacities. In short, one reason for comparative investigation of color vision is to integrate information about color vision with other biological information so as to develop a richer and more complete picture of the natural histories of the species currently inhabiting this planet.

Another, and perhaps more immediate, reason for such comparative investigation is to provide evidence that could guide the search for the mechanisms underlying color vision. Most physiological, anatomical, and biochemical studies concerned with understanding of the mechanisms for vision must necessarily be carried out on nonhuman species. Although at one time many vision researchers were quite cavalier about generalizing capacities from one species to another (indeed, some still are), it should by now be abundantly clear to everyone that species variations are simply too great to permit uninformed cross-species generalizations. If the target mechanisms are those involved in color vision, then naturally it becomes important to know something about the color vision of the species in which the mechanisms are sought. Thus, an important use of comparative color vision information is to permit those interested in mechanisms a choice of an appropriate subject, as well as a firm data base from which intelligent speculations about mechanisms may be made.

A closely related reason for comparative work rests on the hope that species may be identified whose color vision represents a simplified version of our own, providing thereby an arena in which complications can be avoided and insights more easily derived. The reasoning is the same as that which has led to the extensive comparisons frequently made between normal and defective human color vision, the idea being that the defective system represents some sort of a "scaled down" version of the normal system. Although the success of using simplified, model systems to investigate other biological problems cannot be disputed, there are no a priori reasons to suppose that a species whose color vision capacities are in some sense less than ours should therefore simply have a reduced version of our color vision mechanisms. Indeed, given the complexity of the course of evolution, it might be surprising to find that any nonhuman species possesses a straightforward, scaled-down version of man's color vision.

Finally, this overview of reasons for research would be incomplete if I did not mention that much work on comparative color vision has doubtless been motivated by nothing more sophisticated than a curiosity about which of our fellow creatures share our useful, and frequently delightful, color experiences. Curiosity is (fortunately) still a respectable motivation for scientific inquiry! Furthermore, the results produced from this starting point can be every bit as valid and useful as those derived from any of the other perspectives.

This monograph is organized in the following manner: first, the techniques and goals of comparative color vision are described and evaluated; then some material on the mechanisms believed to account for color vision are reviewed; next, a survey indicating the phylogenetic extent of color vision is presented; and finally, some evolutionary and ecological aspects of color vision are examined. The approach employed is selective and invariably idiosyncratic; I have made no attempt to cover all of the extensive body of work done on any of these aspects of comparative color vision. Some of the ground rules for these selections were as follows: first, work on invertebrate species is not included. The major justification for this large omission rests on the very fundamental differences known to exist between the organization of vertebrate and invertebrate visual systems. For those particularly interested in the latter, it is worth noting that the literature on color vision in invertebrates has been reviewed elsewhere (Autrum & Thomas, 1973). Among the vertebrates, I have concentrated on a relatively small number of species. Many species are omitted simply because there are no relevant studies. Beyond that, I have included both those species on which informative behavioral research has been carried out, and those animals that are particularly interesting from the viewpoint of work on color vision mechanisms. Fortunately, and certainly not coincidentally, these two criteria often isolated the same species.

Chapter 2

Techniques and Goals of the Comparative Study of Color Vision

More than 35 years ago Walls (1942) published a comprehensive, provocative, and influential treatise on the biology of the vertebrate eye. The strong views expressed in this remarkable book have been, variously, a source of inspiration and irritation for subsequent students of comparative vision. Among many other topics, Walls subjected research on comparative color vision to an extensive review, noting that "the techniques, some of them very ingenious and some of them very stupid which have been used to ascertain whether particular animals discriminate hue, defy classification." Today it can be concluded that the classification problem no longer exists, if indeed it ever did, and that there has been a virtual disappearance of those techniques that Walls judged so harshly.

I. The "Brightness" Problem

The major difficulty in many of the early studies reviewed by Walls was the inability to ascertain whether differences in brightness or luminance[1] were being

[1]Brightness has been formally defined as "the dimension of color that is referred to a scale of perceptions reporting a color's similarity to some one of a series of achromatic colors ranging from very dim (dark) to very bright (dazzling)" (Burnham, Hanes, & Bartleson, 1963). Thus, brightness is considered to be a "psychological" aspect of color, and the clear implication is that brightness must be established through an ordering or scaling by an observer within the confines of a particular stimulus arrangement. On the other hand, luminance has been formally defined as a photometric quantity, a weighting of radiance according to a standard spectral sensitivity function. In the case of the human, such standardized spectral sensitivity functions have been produced under the auspices of the Commission Internationale d'Eclairage (CIE). In the highly artificial viewing situations in which vision is frequently investigated in laboratories, brightness parallels luminance; that is, the luminance of a pair of featureless, extended surfaces predicts their relative brightnesses. However, in most normal viewing situations brightness does not change coordinately with changes in luminance (for

5

used by the animal as a cue in color discrimination tasks. In fact, many of the early studies made the completely unwarranted assumption that spectral stimuli adjusted to be equally bright for the experimenter would also be equally bright for the animal being tested. Once this problem was recognized and publicized, two strategies emerged to handle it. One method attempted to make brightness an irrelevant cue. This has usually been accomplished by randomly varying the relative luminances of the target stimuli over a wide range, assuming that consistently successful discrimination among the stimuli could then only be achieved on the basis of perceived color differences. For example, an animal might be tested to see if it could discriminate between a chromatic stimulus having a wavelength of 600 nm and an achromatic stimulus with the same spatiotemporal configuration. On successive trials the achromatic stimulus would be varied over a sufficiently wide range of luminances so as to try to assure that it would be brighter on some trials, dimmer on some, and equal in brightness to the 600 nm light on still other trials.

Although logically acceptable, there are two potential problems with this solution. One is that, in fact, on almost all trials there will be a brightness difference between the two stimuli. An animal with only very modest color vision capacities but with exceptionally acute brightness vision might respond by discriminating on the basis of brightness rather than color, for example, by persistently selecting the dimmer of the two stimuli. This technique assumes that brightness differences are irrelevant to the animal, which may not be the case, and thus negative results (that is, a failure to show successful discrimination) must be interpreted cautiously.

A second potential difficulty with the "randomization solution" to the brightness problem concerns the size of the luminance steps employed and their order of presentation. Obviously, if one stimulus is always dimmer than the other the animal may discriminate correctly purely on that basis. In fact, if even a preponderance of negative stimuli are dimmer (or brighter) than the positive stimulus, the animal might still choose solely on the basis of brightness, the resulting performance giving the misleading impression that color vision has been demonstrated. Figure 2.1a illustrates diagrammatically how this might occur in a situation in which the positive light is dimmer than most of the negative lights.

---

examples, see Hurvich & Jameson, 1966). For our present purposes, one of the most important of these occurs when two equiluminant chromatic stimuli are placed side by side; then the more saturated of the two stimuli appears brighter (Boynton, 1973). Obviously, in color-vision studies with nonhuman subjects, equating stimuli to be equiluminant may or may not also make them equally bright, and vice versa. Most studies have not been explicit about whether brightness, luminance, or both were at stake. Since "brightness" has been most frequently referred to in this literature, that term will be frequently used, even in instances where it may not be strictly correct. How to define luminance for nonhuman subjects where the CIE spectral luminosity functions are inappropriate is considered presently.

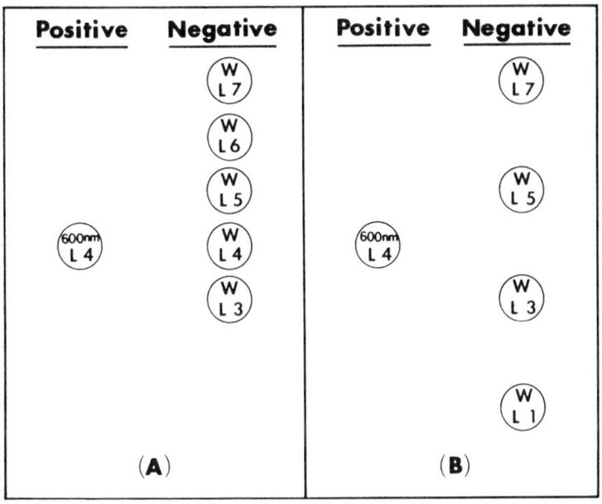

Positive	Negative	Positive	Negative

Figure 2.1 Two possible sources of artifact in a color discrimination test. Each circle repre-sents a stimulus. (W = achromatic light; L = relative luminance). (A) In this test the animal must discriminate the 600-nm light from each of several achromatic lights. Since the positive stimulus has a lower luminance value than a majority of the negative stimuli, the animal could possibly achieve significant levels of correct discrimination by systematically choosing the "dimmer" of each of the stimulus pairs. (B) The negative stimuli all have higher or lower luminance values than does the positive stimulus. Accordingly, it would be possible for the animal to learn an absolute brightness discrimination, that is, solve the problem by always avoiding the light that is brighter than or dimmer than the positive light.

In order to avoid this problem, then, the range of luminance variation should be symmetric about the point of equal brightness. In order to do that, however, the experimenter must know ahead of time at what settings the two stimuli are equally bright! The size of the luminance steps employed also must not be too great. If they are, there are likely to be no occasions on which the two stimuli are equally bright. If so, the animal could theoretically solve the problem by making absolute brightness judgments, by always avoiding the stimulus that is "brighter than" or "dimmer than" the positive stimulus (see Figure 2.1B). Finally, the order of stimulus presentation may also be important. For example, if trials in which the negative stimulus was brighter (or dimmer) than the positive stimulus were presented in, say, blocks of 10 trials, then the animal, using brightness cues alone, could discriminate the positive stimulus correctly on the average 95% of the time by simply learning on the first trial of each block whether to choose the brighter or the dimmer stimulus throughout that series.

All of these behaviors may sound like unreasonably sophisticated strategies to be expected from, for example, a box turtle. However, in conducting experi-ments of this type, two things should be understood. One is that motivated animal subjects will solve discrimination problems any way they can, not necessarily just along the lines planned by the experimenter. Almost everyone who has attempted to test color vision in nonhuman subjects has reached this conclusion as a result of bitter experience. A second fact to be kept in mind is that the burden of proving the interpretation of the results rests, as usual, with the experimenter.

Kicliter and Loop (1976) have suggested one alternative to the practice of varying luminances randomly over a large range. They describe a test situation in which the experimenter first measures the spectral energy distributions of a pair of wide-band spectral stimuli. The animal is then tested to see if it can discriminate between this pair, first when one of the stimuli has a higher absolute radiance at all spectral points and then again when the other has a higher radiance value at all spectral points (see Figure 2.2).

The only assumption made in this test is that brightness increases monotonically with increases in light intensity. One stimulus will be brighter in the first instance whereas the other light will be brighter in the second instance so that a consistency of discrimination indicates that the animal is making a color, not a brightness, discrimination. One difficulty with this technique is that a negative result—that is, a failure to find consistently successful discrimination—may be hard to interpret. Because brightness differences of unknown magnitude are always present in these discriminations, the subject must be forced to ignore them. As was noted earlier, a failure to disregard brightness differences can easily occur in a subject with limited color vision. Furthermore, this test is not easily adapted to determining any details about the subject's color vision. Nevertheless, as Kicliter and Loop (1976) point out, this procedure may be useful in establishing the presence of color vision in cases in which that information alone might be of value.

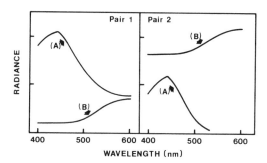

Figure 2.2 Illustration of the spectral radiances of a pair of stimuli (A and B) used in a color discrimination test. In this situation, the animal is first tested on Pair 1 to determine if it can discriminate between the two stimuli when stimulus A has a higher radiance at all spectral wavelengths than stimulus B. In Pair 2 the radiance relationship is reversed with B having a higher radiance at each spectral point. Consistent success at both of these discriminations implies that the performance is based on color not brightness cues.

The second and currently most popular strategy for solving the brightness problem is the positive one. In this case the experimenter makes the to-be-discriminated stimuli equal in effectiveness for the animal. This is done by first ascertaining the animal's sensitivity to various spectral stimuli, usually to the extent of determining a complete spectral sensitivity function. With this information in hand the experimenter can then set the intensities of the test stimuli so as to make them equally luminant for the subject. For example, if the animal's measured threshold to a 500-nm light is 0.7 log units lower than to a 600-nm light, then the 500-nm light must be attenuated by 0.7 log units in order to equate them for effectiveness. Once such equations are made, the color vision tests may proceed. Although considerably more experimental work is involved, in that both sensitivity and color vision tests must be run, this method for handling potential brightness cues seems in many ways much more satisfactory than the techniques that seek to make brightness an irrelevant cue.

Even though the direct approach to brightness control is better, caution is still required. For one thing, given those inevitable fluctuations in threshold that characterize all psychophysical measurements, it is impossible to establish a perfectly precise brightness equation using this technique. Thus, one must also provide some room for error in the brightness equation. This is usually done by varying luminance randomly over a small range around the point of the calculated equation. Note that this procedure then becomes similar to the one employing large luminance variations described previously. The difference is that in this case the experimenter knows where the midpoint (that is, the presumed brightness equation) is located, and so consequently only a relatively small variation in luminance is required.

Some experimental results that clearly illustrate the need for careful control of brightness in comparative studies of color vision are shown in Figure 2.3. The results obtained from testing a prairie dog (*Cynomys ludovicanus*) on a wavelength discrimination task are plotted in this figure. (The nature of color vision in this species is described in Chapter 5.) The animal was tested to see if it could discriminate between two wavelengths that differed by 35 nm. Figure 2.3A shows that when these two monochromatic lights were equated in luminance, the animal was completely unable to discriminate between them (chance performance in this test was 33% correct), even after a large number of test trials. At that point a systematic luminance difference between the two spectral lights was purposely introduced by increasing the luminance of one of them by 0.2 log units. With this relatively small luminance difference the animal immediately began to show successful discrimination, as indicated in Figure 2.3B. The obvious point of this example is that even a small luminance mismatch can lead to the erroneous conclusion that a subject is showing successful color discrimination when, in fact, the discrimination is based on the utilization of a consistent luminance cue.

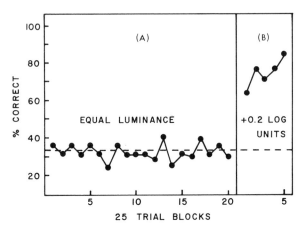

Figure 2.3   Results obtained from a prairie dog (*Cynomys ludovicanus*) tested to determine if it could discriminate between monochromatic stimuli that were either (A) equal in luminance or (B) systematically different in luminance. Chance performance in this test was 33% correct (horizontal dashed line). See the text for further discussion.

Assuming that one uses the direct approach to the luminance equation, there is another basic issue to be considered—the question of how to measure spectral sensitivity. It has been clear for some time that different methods of measuring spectral sensitivity yield quite different spectral sensitivity functions. Specifically, measurements made by flicker photometry or minimum border settings yield curves that approximate the CIE photopic spectral sensitivity function. By comparison, direct brightness matching gives somewhat higher values at the spectral extremes whereas increment-threshold measurements are even more elevated at the spectral extremes (Wagner & Boynton, 1972). The magnitude of these differences is nontrivial, as Figure 2.4 illustrates. Given these differences, it is obvious that the luminance equations made in a color test will depend on the method used to measure spectral sensitivity. As Figure 2.4 shows, this will be particularly critical for those stimuli located near the spectral extremes.

Unfortunately, making a choice among methods to assess spectral sensitivity involves a theoretical judgment, at least to some extent. Many believe that those methods of measurement that yield high sensitivity at the spectral extremes (and often subminima in the middle wavelengths) tap contributions made by neurons carrying both chromatic and achromatic information (Guth, Donley, & Marrocco, 1969). If this is correct, and if the purpose of a luminance equation is to null out the contributions of the achromatic signals only, then the technique of choice should be to use a sensitivity measurement that indexes only that type of signal, as it is argued the flicker and minimum border measurements appear to do.

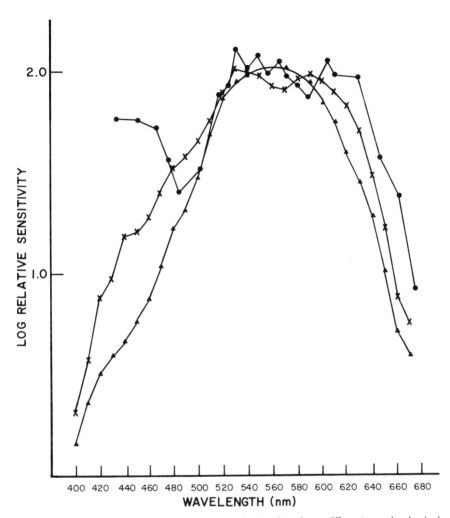

Figure 2.4 Spectral sensitivity functions obtained using three different psychophysical techniques: increment threshold (●) (after Sidley & Sperling, 1967); direct brightness matching (×); and flicker (▲) (after Wagner & Boynton, 1972). The three functions have been arbitrarily equated at 520 nm.

Although the reader may feel that the range of issues to be considered is still an impressive one, there is no question that awareness of and concern about the brightness problem has considerably improved the quality of the comparative color vision literature since Walls initially leveled his devastating judgment.

II. Techniques for Measuring Color Vision

Many experimental procedures have been employed to assess color vision in nonhuman species. Some of these are straightforward adaptations of techniques developed to measure other aspects of animal behavior, while others have been developed specifically for work with a particular species and/or problem. Detailed methodological comments on animal psychophysics have been made elsewhere (Stebbins, 1970; Blough & Yager, 1972). Here I shall cite only a few examples to illustrate the nature of the procedures that have been used to study color vision. In general, these methods may be conveniently divided into those using either learned or unlearned responses.

A. Unlearned Responses

The advantage of using unlearned responses is that no preliminary training is required prior to actual color vision testing. This is a particular advantage when studying those species that are difficult to train or those whose behavioral repertoire may be limited.

Several kinds of unlearned responses have been utilized. The most popular is some form of an optomotor response. These are movements of the eyes, the head, or of the whole body in response to movement in the visual field. The underlying idea is that in order to initiate the motor response, the animal must be able to discriminate the components of the moving stimulus. One commonly used version of this technique involves measuring the occurrence of involuntary eye movements when a patterned stimulus is moved past the animal. Typically this has been accomplished by placing the animal at the center of a rotating drum. The drum surface is patterned, usually striped. The characteristics of the stripes are varied—for example, by changing the contrast between adjacent stripes—until the subject no longer shows following eye movements, which are termed optokinetic nystagmus (OKN). The stripe contrast at which OKN disappears defines threshold sensitivity.

A contemporary version of this test has been described by Wallman (1975). The apparatus is illustrated in Figure 2.5. In this test, the subject views two superimposed grating patterns, which are rotated in opposite directions. The animal's ocular pursuit movements follow the movement of the more conspicuous grating. If the contrast of one of the grating patterns is varied, then the contrast at which the directional pursuit movement disappears defines the point of indiscriminability between the two grating patterns. For example, to test color vision, one grating pattern would be composed of stripes of alternating colors, say red and green, with the luminance contrast between the two stripes variable. The other grating pattern would have similar stripes that are achromatic. If the subject is unable to discriminate between the two colors except on the

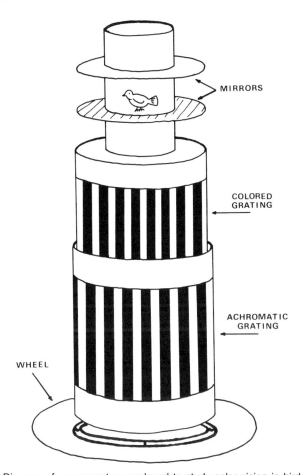

Figure 2.5   Diagram of an apparatus employed to study color vision in birds through the utilization of an optomotor response. The subject is placed inside of two concentrically positioned drums which are then rotated in opposite directions. These drums contain different grating patterns as indicated. The drums are transilluminated by projectors (not shown) located outside of the drums. Details of the measurement procedure are given in the text. (Drawing adapted from Wallman, 1975.)

basis of their relative brightnesses, then there should be some luminance ratio for the red and green stripes that exactly matches the condition of a fixed contrast in the achromatic pattern and thus leads to a loss of the OKN response. Conversely, if color vision is present, the OKN response should persist irrespective of the value of the red to green luminance ratio.

There are both technical and substantive difficulties in using the optomotor responses for testing color vision. First, some means must be employed to force the animal to view the stimulus pattern. This is not as trivial a problem as it may sound, particularly with untamed and often agitated subjects. In fact, the absence of an OKN response has sometimes been attributed to a "lack of interest" on the part of the animal being tested (Suthers, 1966). Second, the width of the stripes on the drum as well as their rate of movement must both be set so they are well within the range to which the subject is sensitive. Finally, and potentially even more problematic, is the possibility that OKN may be suppressed by changes in ocular accommodation. It has been noted that human observers can apparently relax accommodation (that is, "look through" a moving display) and eliminate their OKN. Whether or not other animals can also relax accommodation under these circumstances is not known, but it may deserve some consideration in the interpretation of negative results.

Another difficult, substantive problem in using optomotor responses to measure color vision (or any other perceptual capacity) is that the neural centers involved in controlling the optomotor responses may very well not receive the same sets of retinal information as do those centers involved in the control of discriminative responses. For instance, there is considerable evidence that in the cat visual system, different ganglion cell types, with quite different response properties and quite different retinal distributions, project to the superior colliculus and to the lateral geniculate nucleus (LGN) (Stone & Hoffman, 1971). Similar evidence has been obtained from studies on the ground squirrel visual system (Michael, 1968). Since these different neural centers receive quite different projections and subserve different capacities, it is apparent that tapping the outputs from one may very well not be equivalent to doing so for the other. It is thus conceivable that one might find positive evidence for color vision using a learned response but not using an optomotor response, or vice versa.

Other unlearned responses used to test color vision include a variety of instances in which animals preferentially display some motor behavior with regard to a particular photic display. The notion is that the subject must be able to discriminate the presence of some features of this stimulus in order to initiate the response. This kind of situation has been extensively exploited to study color vision in insects, many of which display movement responses toward objects of certain colors (Autrum & Thomas, 1973). Vertebrates have also been tested using this technique. For example, Muntz (1962) found that frogs (*Rana temporaria*) show a positive phototactic response in that they will preferentially jump toward lights having a predominant short-wavelength composition. The apparatus used to assess preferences of this sort is illustrated in Figure 2.6. Obviously, the brightness problem is as critical here as in other test situations and must be accorded the same attention.

A particularly interesting example of the use of an unlearned preference behavior to study color vision comes from recent work on human infants. Human

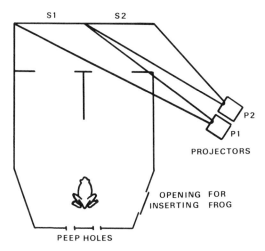

S1 S2

P2

P1

PROJECTORS

OPENING FOR
INSERTING FROG

PEEP HOLES

Figure 2.6 Schematic diagram of an apparatus used to test visual preferences in frogs. The two stimulus panels (S1 and S2) are separately illuminated by projectors (P1 and P2) mounted above the apparatus. The subject is forced to jump toward the panels and any consistencies in the choice of the panel toward which it jumps are recorded. (Adapted from Muntz, 1962.)

infants, when given the option, will stare at patterned visual displays in preference over similar displays that are unpatterned (Fantz, Ordy, & Udelf, 1962). Because this is a reliable behavior, it is possible to utilize this response to study color vision. Thus, Peeples and Teller (1975) tested color vision in 2-month-old human infants by noting which of two stimuli the infant stared at. One stimulus was a square-wave grating pattern with the contrast between adjacent bars variable, while the other was a diffuse field of the same mean luminance level. When the grating pattern was achromatic, the infants preferentially looked at it unless the luminance contrast in the pattern was close to zero. On the other hand, if the grating pattern was composed of red and white bars, the infants chose to stare at the red and white pattern in preference to the unpatterned comparison stimulus irrespective of any relative luminance variations in the red and white bars. It is unlikely that this result could occur unless the subject was capable of some color vision and, thus, this experiment shows in a particularly clear-cut way how an apparently inborn preference may be used to detect the presence of color vision in a subject whose range of behavioral responses is somewhat limited.[2]

[2]To hedge a bit: It is theoretically possible that chromatic-aberration effects induced by the bar pattern might yield some residual brightness variations in the perception of the bar display irrespective of the luminance contrast settings.

B. Learned Responses

A large variety of learned responses have been used to test color vision in nonhuman species. One advantage of the learned response is that the behavior may more closely approximate that involved in visual perception than do the various unlearned responses. Another, perhaps more powerful, advantage is that the experimenter can exercise a greater degree of control over a learned response because the test situation can be structured in a manner deemed desirable by the experimenter, not in the way demanded by some characteristics of the unlearned response. Furthermore, with unlearned responses there is always the real possibility of habituation, that is, of a loss of the behavior following repeated stimulus presentations. For all of these reasons many color vision experimenters have chosen to use some form of a learned response. This is particularly true in those investigations involving vertebrate subjects. A discussion of the motivational and reinforcement issues in experiments of these kinds is available in other sources (for example, Stebbins, 1970; Blough & Yager, 1972). Learned responses are conventionally divided into those based on classical conditioning and those based on instrumental conditioning.

Actually, relatively few experiments on color vision have used classical conditioning paradigms. A good example of the use of classical conditioning techniques (although this particular study was not investigating color vision) is an experiment by Yager and Sharma (1975). In their experiment a goldfish was restrained in a plastic envelope in such a way that lights could be imaged into the eye. Respiration rate was monitored by recording transient temperature changes just in front of the mouth of the fish. Delivery of an electric shock (the unconditioned stimulus) caused a deceleration in respiration rate. A photic stimulus (the conditioned stimulus) was paired with the shock and after a few such paired presentations it began to elicit the conditioned response. Obviously, if the animal cannot see the conditioned stimulus, it will not give a conditioned response. A somewhat analogous procedure making use of classically conditioned changes in heart rate has also been employed to measure visual discrimination in the goldfish (Beauchamp & Rowe, 1977). One arrangement for making such measurements is shown in Figure 2.7. Classical conditioning techniques like these appear most useful for studying species which for various reasons may be difficult to train in an instrumental conditioning situation.

Of those tests based on instrumentally conditioned responses, clearly the most popular is some variation of a forced-choice discrimination task. The subject is first trained to perform some instrumental response for reinforcement, for example, pressing a lever to obtain food or drink. Once this response is learned the subject is presented with a positive and one or more negative stimuli. A response to the positive stimulus is reinforced. If the positive and negative stimuli are discriminable, the animal will eventually solve the problem; if not, chance per-

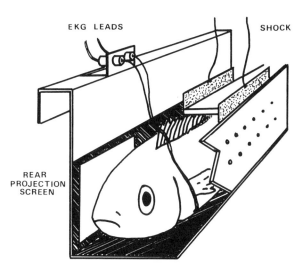

Figure 2.7 Apparatus used to study vision in the fish using a classical conditioning procedure. The fish is completely enclosed in a restraining box. The unconditioned stimulus is an electric shock delivered to the tail. The conditioning stimulus is a light projected onto the screen in front of the eye of the fish. The response measured is a change in heart rate. (Adapted from Powers & Easter, 1978.)

formance will result. By varying the characteristics of the stimuli, the experimenter can test for discrimination between as many stimuli as desired. The advantages of a technique of this kind are, first, that the animal is forced to discriminate between alternative stimuli presented on every trial and, second, that the stimuli intended for discrimination are simultaneously available for comparison. The latter feature typically leads to considerably better discrimination performance than in those cases in which those stimuli intended for discrimination are not simultaneously present.

A famous version of such a forced-choice test, the so-called jumping stand, was developed by Lashley (1938). In his experiments on visual discrimination in rats, the animals were forced to jump from a small platform toward one of two stimulus panels, one of which led to reinforcement (usually food) while the other led to nonreinforcement (in fact, to punishment since the unfortunate animal fell some distance into a net). It is easy to see how color vision tests could be made in this context; indeed, as we shall see in Chapter 5, some of the early studies of rat color vision were carried out using the Lashley jumping stand.

One contemporary version of a forced-choice discrimination apparatus is illustrated in Figure 2.8. In this apparatus, designed for work with monkeys, the animals are trained to touch small transilluminated stimulus panels in order to receive food reinforcement. There are three such panels, each of which is illumi-

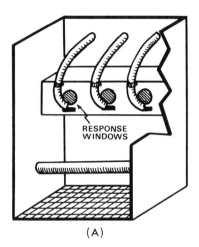

(A)

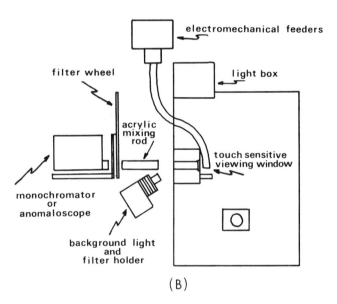

(B)

Figure 2.8 Schematic of a forced-choice discrimination apparatus designed for tests of color vision in monkeys. For details of the test procedure see the text description. (A) The subject's view of the test panels. These are transilluminated from outside of the chamber. If the animal touches the correct panel it is reinforced by delivery of a food pellet into the tray located beneath the window. (B) A side view of the test chamber (to the right) and the optical system used to illuminate the panels. Two light sources are employed, background lights and a monochromator (or an anomaloscope). The latter provides the positive stimulus. Shown also are feeders that are used to deliver reinforcement and a light box that is used to provide chamber illumination.

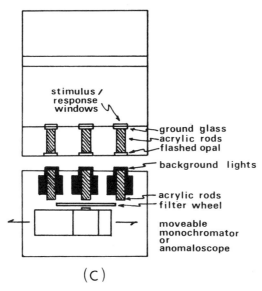

(C)

Figure 2.8 (C) A top view of the optical system. The monochromator (or anomalo-scope) is mounted on a stage which is driven laterally by a motor system so that its output lines up with any one of the three stimulus panels. Each panel can also be illuminated by a background light.

nated from a separate light source. The animal is first trained to press that stimulus panel that appears different from the other two. For example, in a color vision experiment, two of the panels might be illuminated with identical achromatic lights while the third is illuminated with chromatic light (produced by the monochromator illustrated in Figure 2.8). The animal is reinforced if it touches the panel illuminated with chromatic light. Obviously, if the chromatic and achromatic panels have been made equally luminant for the subject, the animal must have color vision in order to correctly make the discrimination. As opposed to the two-alternative, forced-choice procedure developed by Lashley, this situation has two obvious advantages. One is that with three stimulus alterna-tives the probability that the animal will receive reinforcement for random re-sponding is reduced (33% versus 50%). Not only are fewer test trials therefore required, but the discrimination performance often improves as the animal con-tinues to work on hard problems rather than settle for the lower rate of reinforce-ment to be gained by chance levels of performance. A second advantage is that animals can solve the problem on an oddity basis; that is, they can learn to merely select the stimulus that appears different from the others, without regard to the particular characteristics of the stimuli. Because of this, animals may be shifted from one problem to another without extensive retraining because they have not acquired an absolute discrimination. As opposed to this, for correct discrimination in a two-choice situation an animal must both be able to discriminate between the two stimuli and remember which of the two is the correct one. In a three (or more) choice situation the need for learning and remembering which is the correct stimulus is eliminated because the animal need only choose the *different* stimulus, whatever it might be.

Several other forced-choice procedures have also been employed to study color vision in nonhuman subjects. For instance, pigeons have been trained to report whether a pair of visual stimuli appear "the same" or "different." Figure 2.9 illustrates the stimulus arrangement employed in such a test. The birds are first trained to peck on the left and right response keys to obtain food reinforcement. As Figure 2.9 indicates, the center key is made up of a split field, each half of which can be separately transilluminated from light sources located outside of the test chamber. Through an elaborate shaping procedure the birds are trained to peck on, say, the left key if the two halves of the center key appear the same and on the right key if the two appear different. Independent manipulation of the wavelength and intensity of the two center fields permits the experimenter to conduct a large range of color vision experiments in this test situation.

There are many other instrumental conditioning paradigms that have been adapted to test various aspects of visual capacity. Primarily these involve the

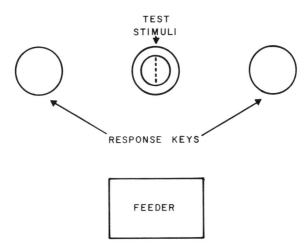

Figure 2.9 Schematic representation of the stimulus panel used in studying color vision in the pigeon. The bird is trained to peck on one of the response keys in order to obtain a brief access to the feeder. The test stimuli (for instance, two monochromatic lights) are projected onto the bipartite field shown on the center key. The bird is trained to indicate whether the two halves of this field appear the same or different by pecking on either one or the other of these keys.

presentation of a single stimulus on each test trial with the subject required to indicate in some manner if (yes–no) or when (temporal forced-choice) the stimulus is detectable. These procedures are particularly appropriate for the measurements of thresholds and thus have been widely used to assess spectral sensitivity, dark adaptation, etc. However, because color vision tests by and large demand not a discrimination of the presence or absence of a stimulus, but rather a concurrent discrimination between two or more stimuli, these single-stimulus techniques have not so far been widely employed in the study of color vision. On the other hand, the forced-choice discrimination tests have been adapted to test color vision in many species of animals as we shall see in Chapters 4 and 5.

III. Appropriate Color Vision Indices

It is a tedious task to assess the color vision of nonhuman species. Usually many weeks or even months must be devoted to obtaining reliable results from a single subject on a single problem. For example, in a study of hue discrimination in the pigeon, Wright (1972b) notes that his subjects were tested for about 750 daily sessions. Because of the effort involved, one might suppose that experimenters would give very careful consideration to the kind of information they want to obtain, but this does not always appear to have been the case. At the least, an experimenter should consider in advance which results about color vision will be most informative, both for purposes of general description of the sensory capacity and as a basis for inferring underlying biological mechanisms.

Obviously, the basic question to ask is whether or not a species has color vision. Indeed, this was the only question asked in many early experiments. There may have once been some justification for stopping with this question, but we now know that the possession of color vision is so widespread that merely establishing its presence in one more species is only of modest interest. The exception to this might be in finding color vision in a species where it seems unlikely (for example, in a cave fish) or establishing its absence in a species where that seems unlikely (for example, the orangutan). In any case, it is inefficient to stop at the point where the presence or absence of color vision has been established because by that time the experimenter has done much of the hard work necessary to move on to more interesting questions about color vision.

Beyond the question of its presence, a more general way to look at color vision is to inquire about its dimensionality. Normal human color vision is characterized as being trichromatic because light from only three widely separated spectral locations can be algebraically added in varying proportions to match the appearance of any other chromatic stimulus. If only two chromatic stimuli are required to complete the matches then the individual is classified as dichromatic. An individual with no color vision is categorized as a monochromat because only the intensity of any single wavelength needs to be adjusted in order to make it match the appearance of any other chromatic stimulus. The basic question of the presence of color vision can thus be seen as an attempt to find out if the visual system can be characterized as spectrally one-dimensional, or whether it is more than one-dimensional. This question can obviously (and usefully) be expanded to ask how many dimensions are required: is it dichromatic, trichromatic, or perhaps tetrachromatic?

A straightforward approach to the question of the dimensionality follows directly from the basic test for the presence of color vison. If that test has been done in the usual way, that is, by requiring the subject to discriminate between equiluminant monochromatic and achromatic lights, then a simple extension of the same test to include a wide range of monochromatic lights of different wavelengths will show if the subject has dichromatic or trichromatic vision. If the subject is dichromatic, there must be some band of wavelengths that is indiscriminable from an equiluminant achromatic light. This spectral location is called the neutral point. If the animal has no spectral neutral point then its color vision must be based on (at least) three spectral mechanisms. Naturally, some caution is required in interpreting the results of this test because the spectral width of the neutral point is often quite restricted (Walls & Heath, 1956), perhaps no more than a few nanometers, and thus could be easily missed entirely with an insensitive test. If a neutral point is found, its spectral location may be diagnostic as to the nature of the animal's color vision (see Section IV-B).

Beyond knowing the dimensionality of the color vision, information about the spectral sensitivities of the cone receptors and how their outputs interact is also useful in a description of color vision. One can attempt to estimate the spectral absorption characteristics of the various receptors psychophysically by measuring the spectral sensitivity of the animal both with and without concurrent chromatic adaptation. Such measurements made in man (Stiles, 1949) appear to yield only an approximation (but see Bowmaker & Dartnall, 1980) to the cone spectral absorption curves, but greater success might be expected in species possessing more widely separated cone absorption functions.

It is also useful to test color mixing. One common instance of this is the Rayleigh match—determining what relative mixture of red and green lights are indiscriminable from a monochromatic yellow light. Since this test clearly differentiates humans with normal color vision from those with the two most common types of trichromatic anomalies (see Section IV-2), it is an important index of color vision in other animals as well. Although it has not often been done, a much broader range of color matches than those necessary for specifying the Rayleigh match could also be carried out on nonhumans.

The color vision tests described give information principally about the initial stages of processing in the visual system. Indeed, color mixture data are conventionally interpreted in terms of the relative spectral sensitivies of the cones. Obviously, however, many of the interesting aspects of color vision involve processing at neural levels beyond the photoreceptors. Tests that tap these ''higher'' levels are useful adjuncts to the battery of color vision tests. Furthermore, the color vision tests described give no real indication of the degree of color vision possessed by an animal. It is important to appreciate that color vision can vary not only in type (for example, dichromatic or trichromatic) but also in *amount*. Two individuals could have ''normal'' trichromatic color vision but one might still be able to discern much smaller color differences than the other. This aspect of color vision is important because the acuteness of color vision for a particular species may directly suggest how important the capacity of color vision is for that species.

Of the possible indices of the acuteness of color vision, wavelength and colorimetric purity discriminations have been the most widely tested. Both the shape of the wavelength discrimination function and the absolute size of the discrimination steps are of importance, the shape for the indications it gives about the nature of the interactions between spectral mechanisms, and the absolute size of the step as an indication of the amount of color vision the animal possesses. Similarly, measures of colorimetric purity discrimination are important for providing some basis for speculation about the relationship between achromatic and chromatic processing mechanisms, and for indicating just how ''colored'' the spectrum appears to the animal being tested.

To summarize, it seems important to do more than just establish the presence of color vision in a species. To provide a useful description of its color vision one should measure other aspects of this capacity: the number of chromatic dimensions, the spectral characteristics of the receptors and the nature of the interactions among their outputs, and the extent of color vision which the animal possesses. All of these can be derived directly or inferred from appropriate psychophysical measurements. Finally, although not yet a common procedure in comparative studies, the growing interest in the spatial and temporal properties of color vision, as well as the interactions between these dimensions, suggests that a variety of tests designed to tap these more dynamic properties of vision might be useful future additions to the basic color vision tests.

IV. Some Basic Data on Human Color Vision

The vast majority of the information about color vision comes from studies done on humans. It is therefore hardly surprising that results from other species are almost always interpreted by the experimenter in the context of human color vision. In the previous section I suggested some of the kinds of information about color vision that can be considered descriptively useful. Because color vision results are often treated comparatively with human data, it may be worthwhile to review very briefly some of the basic aspects of color vision in humans. Accordingly, what follows is a short and highly selective summary of some features of human color vision. A recent, very extensive treatment of this topic has been provided by Boynton (1979).

A. The Normal Trichromat

As already indicated, the human norm is a three-variable system. The trichromat is able to match precisely the appearance of any other chromatic stimulus by algebraically mixing three widely separated spectral wavelengths. Furthermore, given a standardized situation in which the stimuli and other features of the test arrangement are kept constant, the proportions in the mixture chosen by the normal trichromat do not vary greatly from individual to individual. For instance, Figure 2.10 shows the relative proportions of 650-, 530-, and 460-nm light which several trichromats mixed together in order to match the appearances of various monochromatic lights covering the spectrum from 400 to 700 nm. The mixture functions appear much as one might expect. Thus, for example, when the stimulus whose appearance is to be matched is of long wavelength, then a large proportion of the mixture must be made up of the long-wavelength component (650 nm). Similar considerations hold for other test stimuli. The fact that the proportion of one of the mixture components sometimes has a negative value

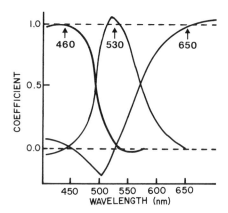

Figure 2.10 Color mixture coefficients for normal human trichromats. The three curves give the relative proportions of three monochromatic lights (650, 530, and 460 nm) that human observers mixed together in order to match the appearances of those spectral stimuli whose values are shown on the abscissa. (Data replotted from Wright, 1947.)

means simply that this component was moved to the side of the test field containing the spectral light the subject was trying to match. Although color-mixing experiments are at the very heart of the study of human color vision, only a limited number of such experiments have so far been done with nonhuman subjects.

Probably the most common referent for color vision is wavelength discrimination. Figure 2.11 shows the results from several different wavelength discrimination studies carried out on normal human trichromats. Each point represents the magnitude of the wavelength difference needed for discrimination at each of the wavelength values shown on the abscissa. The characteristic shape of these functions indicate that there are two regions of most acute wavelength discrimination, just short of 500 nm and at about 580 to 590 nm. At the same time, there is relatively poor discrimination through the middle wavelengths and then very poor discrimination at the spectral extremes. A second feature worth noting is just how good normal human trichromats are at the wavelength discrimination task—wavelength differences of less than 1 nm can be successfully discriminated at a number of spectral locations. In the context of earlier discussion, the normal human trichromat clearly has exceedingly acute color vision.

A second commonly used referent in comparative color vision involves an investigation of the effects of the colorimetric purity of a spectral stimulus on its perceived color. Although this kind of dependence has been measured in many

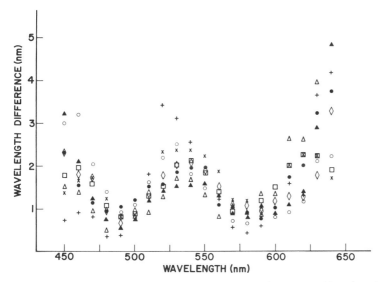

Figure 2.11   Thresholds for wavelength discrimination at various spectral locations for normal human trichromats. Each symbol represents the results from one study. (Data replotted from Wright, 1947.)

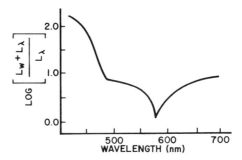

Figure 2.12   Spectral saturation for the normal human trichromat. The curve plots the amount of achromatic light (L_w) that had to be added to a monochromatic light (L_λ) relative to L_w at various wavelengths in order to render the monochromatic light indiscriminable from other, equally luminant, achromatic lights.

ways, a convenient way to think of it is in terms of how much achromatic light needs to be added to a particular chromatic light in order to render this mixture indiscriminable from an equiluminant achromatic light. In essence it asks how colored a spectral stimulus is perceived to be. Figure 2.12 answers this question for the normal human trichromat. The data shown indicate that the spectral extremes appear highly colored (saturated) to these subjects. Saturation decreases as wavelength is shifted toward the middle of the spectrum, reaching a minimal value at about 570 to 580 nm. Sometimes measurements of this kind also reveal a secondary region of lowered saturation at about 500 nm. These results are in accord with our visual experience—spectral reds and blues appear highly colored whereas yellows do not.

B. Defective Color Vision

This rather imprecise phrase is used to describe those human observers whose color vision differs in significant ways from that of the prototypical normal trichromat. Much interest has been paid to defective color vision, primarily because of the belief that study of the defective case might enhance our understanding of the mechanisms for normal color vision. Defective color vision is characteristically subdivided along the lines of its ontogenetic origin, whether congenital or acquired. Although the acquired color vision defects are described as "common" (Grutzner, 1972), they are also generally less clear-cut in their defining characteristics than the congenital defects, partly because they are often associated with other serious ocular problems. Perhaps because of this they have also been much less intensively studied. An extensive review of both congenital and acquired color vision defects by Pokorny, Smith, Verriest, and Pinckers (1979) is highly recommended for further discussion of these topics.

Table 2.1

Frequency of Defective Color Vision among Humans[a]

Classification	Percentage
Anomalous Trichromacy	(5.6)
Protanomaly	1.0
Deuteranomaly	4.6
Tritanomaly	0.0001
Dichromacy	(2.6)
Protanopia	1.2
Deuteranopia	1.4
Tritanopia	0.008–0.0015
Monochromacy	(0.003)
Total Incidence of Color	
Defective Vision	8.2

[a] From Marriott (1962).

Congenitally derived, defective color vision occurs in about 8% of the members of various white races, whereas it is somewhat less frequent in other racial groups (Jaeger, 1972). Nearly all forms of congenital defective color vision are inherited as sex-linked recessive traits so most of the affected individuals are males (only about 0.4% of all females have defective color vision). Table 2.1 gives one compilation of the relative frequencies of humans in the various categories of defective color vision, each of which will now be briefly described.

1. Anomalous Trichromacy

Like normal trichromats, these individuals have a three-variable color vision system. They differ from the norm in the relative proportions of the mixture components they use to match the appearances of various monochromatic lights. Nearly all of these individuals fall into two groups—protanomalous and deuteranomalous trichromats. The standard measure used to classify these individuals is the previously mentioned Rayleigh match, in which the subject mixes together 535- and 670-nm lights so as to match the appearance of a 589-nm light. The relative proportions of the 535- and 670-nm lights used in the mixture is diagnostic in that all normal trichromats use roughly the same proportions, whereas the protanomalous individuals use more of the 670-nm component and the deuteranomalous subjects use more of the 535-nm component. If the ratio of 670- to 535-nm light used by the subject is expressed relative to the ratio used by a normal trichromat, an index called the anomalous quotient is obtained. Figure 2.13 shows the distributions of anomalous quotients derived from a large sample of human subjects. This figure verifies that (1) all normal trichromats use about

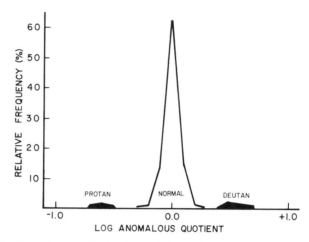

Figure 2.13 Anomalous quotients obtained from a large sample of human observers. Those values derived from the two major groups of anomalous trichromats are shaded. (Adapted from Schmidt, 1955.)

the same ratio of 670- to 535-nm light while the protan and deutan observers require much more of one of the two components, and (2) suggests that the distribution of matches is not continuous, that is, the protan and deutan observers do not appear to simply represent extreme cases of normal trichromacy. The third category of anomalous trichromacy, tritanomaly, is exceedingly rare (see Table 2.1). Like the other anomalous individuals they use abnormal proportions in a trichromatic mixture, in this case much more of the short-wavelength component.

The trichromatic anomalies are considered to be the most difficult of the color defects to provide complete explanations for. This arises partially because these groups are not homogeneous. For example, although the relative proportions of 670- and 535-nm light used in the Rayleigh match are roughly similar across all protanomalous or deuteranomalous observers (see Figure 2.13), there is substantial variation in the precision with which anomalous observers in each group are able to make such matches. That is, for some anomalous observers, approximately the same mixture proportions are chosen each time a match is made, while for others successive matches produce a wide variation in the proportions of 670- and 535-nm light selected (Hurvich, 1972). A similar variation is evident in other indices of color vision obtained from anomalous trichromats. Both the wavelength and purity discrimination functions obtained from anomalous observers differ not only from similar measurements made on normal trichromats, but also show much variation within each category of anomalous observer (Wright, 1947). The extent of such variation in wavelength discrimination among a group of protanomalous humans is summarized in Figure 2.14.

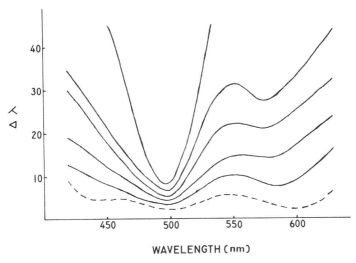

Figure 2.14 Wavelength discrimination curves for normal human trichromats (dashed line) and several different protanomalous observers (solid lines). The data from the protan observers indicate that the severity of loss in wavelength discrimination may be substantially variable among individuals all of whom fall within the same formal classification. (Plotted from measurements made by Wright, 1947.)

2. Dichromacy

As the name implies, unlike trichromats, dichromats require only two components in an additive mixture to match the appearance of all spectral stimuli. These individuals show severely reduced ability to discriminate between certain pairs of wavelengths and their saturation discrimination functions are also substantially different from those of normal trichromats. However, undoubtedly the most dramatic aspect of dichromatic color vision is the occurrence of a spectral neutral point, a restricted region of the spectrum which to the dichromat appears the same as a broad-band (achromatic) light. Neutral points for the very rare tritanopes are in the vicinity of 570 to 580 nm. Neutral-point locations for protanopes and deuteranopes are located closer together, somewhere in the vicinity of 490 to 505 nm. Considerable experimental effort has been directed to specifying the exact locations of these neutral points, and toward trying to decide if neutral-point locations are systematically different in protanopes and deuteranopes, enough so that the spectral location of the neutral point might be used as a diagnostic tool. Unfortunately, there are not yet definitive answers to these questions. In general, protanopic neutral-point locations are centered at shorter wavelengths than deuteranopic neutral points. However, although some investigators have found no overlap in the distribution of spectral neutral point locations for these dichromats (for example, Walls & Heath, 1956), other,

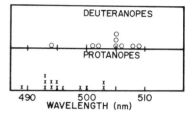

Figure 2.15 Neutral point locations for human dichromats. Each symbol represents the neutral point location for one subject. (Adapted from Hurvich & Jameson, 1974.)

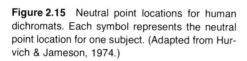

equally carefully done studies produce such overlap. Figure 2.15 shows distributions of neutral points for protanopes and deuteranopes as measured in a study which found an overlap between the two groups. Surely part of the reason why this question has not been settled is because of the wide range of variables that influence measured neutral-point locations: things such as the nature of the achromatic comparison light, the degree of intraocular pigmentation in individual subjects, and the type of behavioral technique employed.

Because a search for a spectral neutral point is often made as a part of an investigation of color vision in nonhuman species, it is worth noting that the test run on animals usually differs radically from that employed to assess neutral points in humans. Animal studies typically use a forced-choice procedure in which the animal is tested to determine if it can discriminate various spectral wavelengths from an equiluminant achromatic light. Human studies typically involve a matching or color-naming procedure. This difference is significant. For the trained animal, if there are *any* perceived differences between the spectral light and the achromatic light the discrimination will always be successfully made, that is, no neutral point will be found. On the other hand, because the human subject is instructed to match the appearances of the spectral and achromatic stimuli, the wavelength selected will be that which appears "most like" the achromatic stimulus. The point is that "matching" and "indiscriminable" are not necessarily equivalent. The discrimination test is considerably more rigorous. Indeed, human subjects who reliably indicate a neutral point location in a matching task will sometimes show little difficulty in discriminating between the two components of their match and thus would be classified as dichromatic by the matching test but not by the discrimination test.[3]

The dichromat is able to discriminate fairly small wavelength differences in the immediate vicinity of the neutral point. However, at wavelengths only

[3]The issue raised here suggests that the boundaries between the various categories of color defective vision are not as discrete as a reader of textbooks is often led to believe, and that the particular details of the testing situation may critically affect the diagnosis reached. There is further experimental evidence to substantiate this view. For example, individuals who appear dichromatic by virtue of their performance on several standard color vision tests are frequently able to set precise Rayleigh matches, a capacity that they should not have if they have dichromatic color vision (Nagy,

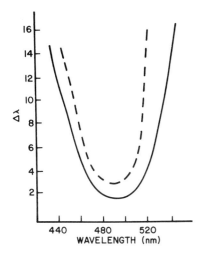

Figure 2.16 Wavelength discrimination for human deuteranopes (—) and protanopes (--). (Adapted from Wright, 1947.)

slightly displaced from this location, discrimination falls off drastically. The result is that the wavelength discrimination functions for human protanopes and deuteranopes are roughly U-shaped (see Figure 2.16), and they are quite similar for the two types of dichromacy—enough so that these measurements cannot be used to differentiate the two types of dichromacy.

3. Monochromacy

Monochromats lack color vision. They are able to match the appearance of two lights by simply adjusting the intensity of one of them. The severity of this color vision defect and the possibility that it might represent the ultimate reduction of the normal color vision system has made this condition an attractive one for study. However, because monochromats are few in number, and because it has turned out there are several distinct sets of capacities subsumed under this single heading, our understanding of monochromacy is still incomplete. Classically it was supposed that there are two types of monochromacy: (1) rod monochromacy in which the retina has either few or no cones in conjunction with a normal rod population, and (2) cone monochromacy in which the photoreceptor complement of the retina was normal but in which there was an apparent gross abnormality of neural connections. It has been shown that within both of these classes several

---

1980). The reason for these disparities in classification is, at least partly, that some individuals truly fall within the boundary regions between the various color vision classes, and whether they are diagnosed as falling on one side or the other depends on whether the color vision test presents their visual systems with a strong or a weak challenge.

different forms of monochromacy exist (Alpern, 1974). To explain the mechanisms responsible for all of these, appeals are made to the full range of logical possibilities, either singly or in combination: missing photopigment classes, missing receptor types, and abnormal neural connections.

V. Cross-Species Standards for Color Vision

A curious feature common to most discussions involving cross-species comparisons of sensory capacity is that of standards. It is usual to find the sensory capacities of nonhuman species characterized through the use of a qualitative descriptor. For example, a species might be described as having "poor," "weak," or "excellent" color vision. What is meant by such labels is typically not made explicit. There are at least two aspects to this ambiguity. First, we have already noted that color vision may vary both qualitatively and quantitatively so the descriptor could be in reference to either of these. Thus, a dichromat could be described as having "poor" color vision, but so too could a "normal" trichromat whose wavelength difference thresholds were unusually large. Secondly, to what standards do these qualifiers apply? In the case of color vision the answer to this question is, clearly, the normal trichromacy of man. Implicit here is the strong belief that man possesses the "best" color vision among the vertebrates. Whether or not this is true in an objective sense is not certain. Simply on the basis of a comparison of wavelength difference thresholds, for example, it would appear that the pigeon at least approaches the levels achieved by humans (see Chapter 4). Beyond a mere species chauvinism, perhaps misplaced, it is possible to argue that a "good" sensory capacity is one which permits an organism to efficiently extract and process that environmental information required in the conduct of its normal behavior. In this view an animal with very limited color vision might indeed be said to have "good" color vision.

Despite the ambiguity of such qualifiers, evaluative comparisons of the color vision of many species are made throughout this work. These are to be taken only as a convenient shorthand. Thus, "poor," "excellent," and so on, are to be understood in reference to the capacities of the average normal human trichromat.

Chapter 3

Mechanisms for Color Vision

As I have noted previously, much of the interest in comparative color vision has arisen from the enhanced possibilities for seeking the mechanisms underlying color vision that is provided by study of nonhuman subjects. Hence, to reasonably discuss comparative color vision, constant reference must be made to the range of biological adaptations accounting for this capacity. The intent of this chapter is to provide some background material on these issues by reviewing in a brief and selective manner what is by now a very large body of literature detailing the mechanisms underlying color vision. To make such a venture maximally useful, references are provided to a number of recent reviews of this literature.

I. Structural Overview of the Eye

Although the ocular structure among all vertebrates is similar, nevertheless, there is considerable variation in the details. This section provides a brief structural orientation; a more detailed discussion on those topics important for color vision will follow in later sections.

Figure 3.1 is a familiar schematic diagram representing a horizontal cut made through a human eye. It illustrates the major structural components of vertebrate eyes. Light enters the eye (from above in Figure 3.1) through the transparent cornea. This element acts as the primary refractive element which, in conjunction with other refractive surfaces, principally the lens, forms an image of the visual scene on the surface of the retina. The tough, opaque sclera constitutes the posterior continuation of the cornea. At the optic nerve head the sclera becomes continuous with the sheath surrounding the optic nerve. Under the action of the ciliary muscles the iris constricts and dilates to form a variable sized opening, the pupil, which influences the amount of light that passes into the lens and onto the retinal surface. The size of the pupillary aperture changes in response to changes in the amount of light entering the eye, as well as being influenced by a wide range of other factors. Changes in pupil size, in addition to regulating grossly the amount of light entering the eye, importantly affects the optical quality of the image formed on the retinal surface. The two large cavities within the eye,

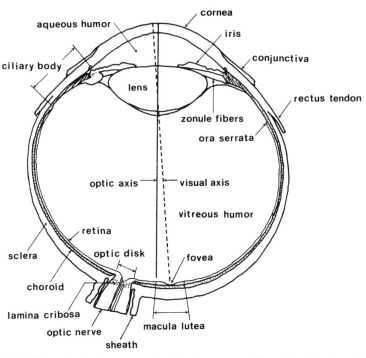

Figure 3.1 Major structural components of the human eye. (From *The Vertebrate Retina: Principles of Structure and Function* by R. W. Rodieck. W. H. Freeman and Company. Copyright © 1973.)

anterior and posterior to the lens, are filled with transparent media, the aqueous and vitreous humor, respectively. The former is continuously circulated so as to maintain a relatively stable intraocular pressure. The choroid contains a dense vascular bed which, most importantly, transports nutrients to the retinal photoreceptors.

The retina is part of the central nervous system. It is a delicate tissue, 200–300 μm in thickness and made up of some 200 million cells in the human, which forms the lining layer of the eyeball. Figure 3.2 is a low-power photomicrograph showing a cross section of the retina. In this orientation, light enters the retina from below, on the vitreal side. The retina is a laminated structure composed of several cell layers separated by zones where these cells interact synaptically. At the top is the choroid and, immediately beneath, a densely pigmented layer—the pigment epithelium. Abutting the pigment epithelium are the distal processes of the photoreceptors, the rods and cones. These cells will be discussed in greater detail presently. The outer nuclear layer is composed of the cell bodies of the

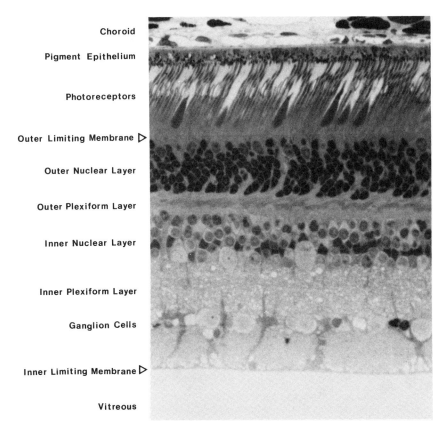

Figure 3.2 Cross-section taken from the parafoveal retina of the rhesus monkey. Magnification: × 397. (Photomicrograph taken by D. H. Anderson.)

photoreceptors. Just below is the outer plexiform layer, a dense synaptic bed where the photoreceptor processes, the processes of the bipolar cells, and those of a third class of retinal cell, the horizontal cells, all interact. The cell bodies of the bipolar cells are located in the inner nuclear layer. The inner plexiform layer is a second zone of complex synaptic interactions, in this case between bipolar cells, ganglion cells, and amacrine cells. The cell bodies located closest to the vitreal surface belong to the ganglion cells. Their axonal processes constitute the optic nerve fibers. They pass over the retinal surface and are gathered together at the optic nerve head to form the second cranial nerve. In the human eye there are about 1 million such fibers.

Although the cell types just listed and their relative retinal locations are common to most vertebrate eyes, it has become clear that in terms of the relative

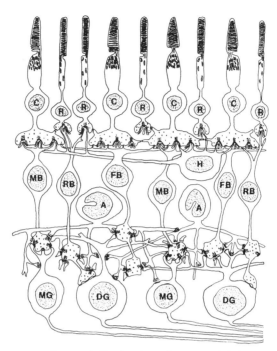

Figure 3.3 A summary diagram indicating the cell types and their interconnections in the primate retina. (R, rod; C, cone; RB, rod bipolar; MB, midget bipolar; FB, flat bipolar; H, horizontal cell; A, amacrine cell; MG, midget ganglion cell; DG, diffuse ganglion cell.) (After Dowling & Boycott, 1966.)

numbers, ultrastructure, and the kinds of synaptic contacts formed, there is considerable variation from species to species. Figure 3.3 shows a schematic cross section of the primate retina as determined by Dowling and Boycott (1966) from a combined consideration of gross structure and electron microscopy. The figure is intended to characterize the types of synaptic contacts; it is not in scale with regard to the relative locations of these retinal cells or to the actual number of the various types of elements. According to their analysis (Dowling & Boycott, 1966), the photoreceptors (R and C) form synaptic contacts with (1) other photoreceptors, (2) horizontal cells, and (3) bipolar cells. The latter two classes of elements form contacts in the specialized terminal regions of the photoreceptors. In primate retinas three classes of bipolar cells are identified: those making contact with rods (RB), and two classes contacting cones, midget (MB) and flat bipolars (FB). The latter two are differentiated by virtue of the structure of the contacts they form with the photoreceptors. In the inner plexiform layer synapses are formed between bipolar, amacrine, and ganglion cells. As indicated schematically, there are numerous potential locations of synaptic interactions among these cells. Finally, two classes of ganglion cells are also distinguished in the primate retina, midget (MG) and diffuse (DG) ganglion cells. As Figure 3.3 shows, the two differ in the degree of their dendritic arborizations with the diffuse ganglion cells receiving inputs over a much wider spatial extent than the midget ganglion cells.

A central feature of retinal construction is that it is spatially heterogeneous: the densities and relative proportions of the various cell types vary continuously across the retinal extent. The best known manifestation of this heterogeneity is in the primate retina. Thus, in a centrally located specialization called the fovea (see Figure 3.1), the retina is greatly thinned because the neural layers of the retina located proximal to the photoreceptors are swept aside laterally. At the same time, the density of one photoreceptor type, the cones, rises to its maximum value, about 150,000 per square millimeter in the human. From the center of the primate fovea, cone density drops off rapidly while at the same time the density of rods increases toward more peripheral locations. The consequence of this heterogeneity in receptor composition, along with equally important changes in the relative proportions of other retinal cells, is that vision varies significantly as a function of the retinal location stimulated.

All vertebrate retinas show some degree of structural heterogeneity of the kind just described for the human. Not surprisingly, however, the nature and magnitude of this heterogeneity varies substantially from species to species. Figure 3.4 illustrates this point. Estimates are shown of rod and cone density as a function of retinal location for three mammalian species on which such measurements have been made. One of them is for man, while the other two are nocturnal species—the domestic cat and the owl monkey (*Aotus trivirgatus*). In those graphs angular displacement is given relative to the center of the fovea for man, and from the midpoint of the *area centralis* for the other species. Note first that rod density for the human follows the pattern previously described, that is, there are no rods in the central fovea, increases in density for more peripheral locations, and a peak density for these receptors located at about 20° from the fovea. The two nocturnal species differ considerably, both from the human and from each other. The cat has a much higher overall rod density with a peak located, as in the human, at about 20° from the center of the retina. Also like the human retina, the cat retina shows a decrease in rod density near the center of the retina although there are still a considerable number of rods even at its very center. The owl monkey retina also has a higher overall rod density than that found in the human. In this case, however, rod density increases rather continuously from the periphery toward the central part of the retina. The measurements of cone density shown in Figure 3.4 also underlines the sharp species differences. In this instance all three species show highest cone density in the central part of the retina. However, the density function is sharply peaked for man and cat while showing only a modest central increase for the owl monkey. The maximum cone

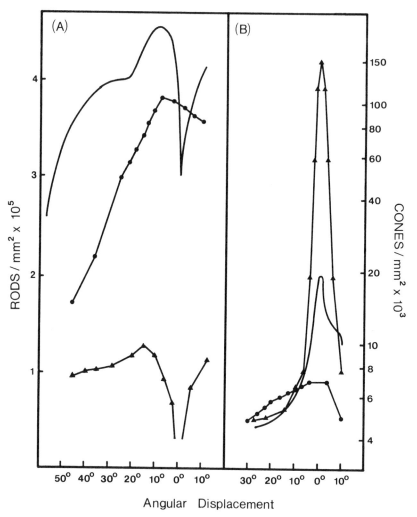

Figure 3.4 Rod and cone densities for three mammalian species: cat (smooth curve); human (▲—▲); owl monkey (●—●). (A) rods; (B) cones. In each panel, values to the left are for the temporal retina, those to the right are for the nasal retina. The counts represent densities along a horizontal meridian. Different scaling is used for plots of the two receptor types because of the relatively low cone densities found in the owl monkey. (After Ogden, 1975.)

density values are greatly different for these three species. For example, cone density is over twenty times as great in the center of the retina in man as compared to the nocturnal owl monkey.

II. Ocular Filters

To be effective as a stimulus for vision, light must reach the photoreceptor outer segments and be absorbed by the photopigments contained therein. As much as 50% of the light passing through the ocular media does not contribute to vision because it is absorbed, reflected, or scattered before it reaches the retina. To the extent that preretinal absorption and reflectance are spectrally selective, that is, are more effective for some wavelengths than others, the properties of the preretinal filters are potentially relevant to an understanding of the mechanisms for color vision in a species. There are two important classes of such ocular filters: absorbance by the lens and cornea, and absorbance by intrareceptor oil droplets. The literature on these classes of filters has been the subject of two comprehensive reviews: Walls (1942) extensively reviewed the early literature, whereas Muntz (1972) has covered the subject for the subsequent 3-decade interval.

A. Absorbance by the Lens

The lenses of many species appear yellow—that is, they preferentially absorb short-wavelength light. This coloring is particularly obvious in those mammals that are strongly diurnal. The spectral filtering by lenses has frequently been assessed by spectrophotometric measurements of excised tissue. As measured in this way, the degree of pigmentation varies significantly across species. Figure 3.5 shows the pattern of spectral absorbance measured in the lenses of several different species. As illustrated, the lenses of all the species show little or no spectrally selective absorbance of light at wavelengths longer than 500 nm. Below 500 nm, absorbance values increase more-or-less rapidly. Among mammals the degree of absorption of light by the lens in the short wavelengths is at least roughly correlated with the degree of diurnality of the animal. Thus for very strongly diurnal animals, like most squirrels, lens density achieves such high values as to make their lenses essentially opaque to short wavelengths. For example, Cooper and Robson (1969) estimated that the lens of the gray squirrel has an optical density of about 9 at 400 nm. On the other hand for nocturnal species, like the flying squirrel (see Figure 3.5), there is only very slight spectrally selective absorption over this same range. In addition to these species differences there are also, at least in man, significant increases in lens density associated with the age of the individual. Individual differences in lens density

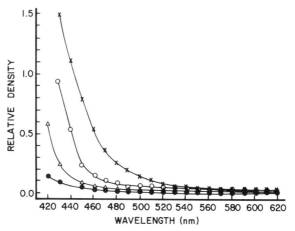

Figure 3.5 Average spectral absorbance by the lenses of several different species: X, ground squirrel; O, tree squirrel; △, monkey; ●, flying squirrel. The ground squirrels are strongly diurnal, the flying squirrel strongly nocturnal. (Measurements on squirrel lenses taken from Yolton, Yolton, Renz & Jacobs, 1974.)

have often been used as part of an explanation for intersubject variations in color vision. A compilation of a considerable amount of data on human lens absorbance is given by Norren and Vos (1974).

The possible adaptive significance of the attenuation of short-wavelength energy by the lens has been the subject of much speculation (see Muntz, 1972). It has been variously suggested that

1. the yellowing of the lens serves to reduce the effects of chromatic aberration imposed by the optics of the eye;
2. the yellow lens serves to reduce the effects of scattering of short-wavelength light and is thus an adaptation to enhance spatial resolution;
3. the lens screens from the retina short-wavelength (ultraviolet) radiation that is potentially injurious, or
4. the lens may reduce the likelihood that the secondary absorbance peaks that all photopigments appear to possess will absorb light and thus trigger visual activation (see the following).

There is no compelling evidence to provide strong support for any of these possibilities, but neither can any of them be easily rejected. Perhaps all of them are important. Of course, it is also possible that spectral filtering by the lens is an unavoidable by-product of lens composition, and that it consequently has no visually adaptive significance. The age-related changes previously referred to make this a possibility, although for many who believe with Walls (1942) that "everything in the visual system means something," this alternative is unattractive. It is at any rate more than a little curious that the lenses in many species serve to specifically attenuate those spectral wavelengths to which their retinal photopigments have very substantial sensitivity.

B. Oil Droplets

Oil droplets are found in the retinas of a variety of different species, including some fish, amphibians, birds, and reptiles. These droplets, basically spherical in shape, are positioned in front of the outer segments of some of the cone photoreceptors found in these retinas. Generally there is only one droplet per receptor (see Figure 3.6). These droplets consist of lipids in which frequently carotenoid pigments are dissolved (Meyer, Cooper, & Gernez, 1965). The latter impart to the droplets a vivid color. The absorption characteristics of the oil droplets have been measured for a number of different species (Muntz, 1972). In essence, the droplets serve as high-pass filters, absorbing strongly and nearly uniformly throughout the short wavelengths and then showing a rapid falloff in absorbance to the long wavelengths. This property is illustrated by the absorbance spectra shown in Figure 3.7. As shown in that figure, the droplets differ in the spectral location where they begin to transmit appreciable amounts of light, and it is this variation that leads to the differences in the colors of the various classes of droplets. Basically, the droplets appear to our eyes as red, orange, yellow, or occasionally, greenish-yellow.

Due to the location of the oil droplet in the photoreceptor, some of the light penetrating to the photopigment contained in the cones will likely be spectrally filtered. It is hardly surprising, therefore, that various theorists have supposed that the oil droplets might be part of a color vision mechanism (see Walls, 1942; Muntz, 1972; Wolbarsht, 1976). Indeed, it has sometimes been seriously proposed that the oil droplets might be *the* pivotal mechanism. The adequacy of this contention and the actual participation of the oil droplets in the color vision process are considered in detail later on in conjunction with an examination of those species where the idea might have some relevance (see Chapter 4). However, it can be pointed out here that the retinal oil droplets quite clearly *do not* constitute the sole mechanism for producing color vision. Indeed, it is to date unclear just what role they have in the color vision process per se.

Even though oil droplets appear not to serve as a primary color vision mechanism, they will certainly affect vision in those species that possess them. The directions of this influence are potentially the same as those ascribed to lens pigmentation. Not surprisingly, therefore, the same kinds of adaptive functions already suggested for lens pigmentation have also been raised for oil droplets. For instance, Wolbarsht (1976) has suggested that the primary purpose of the oil droplets may be to reduce the effectiveness of the secondary absorbance peaks that photopigments have in the short wavelengths, and/or perhaps serve as de-

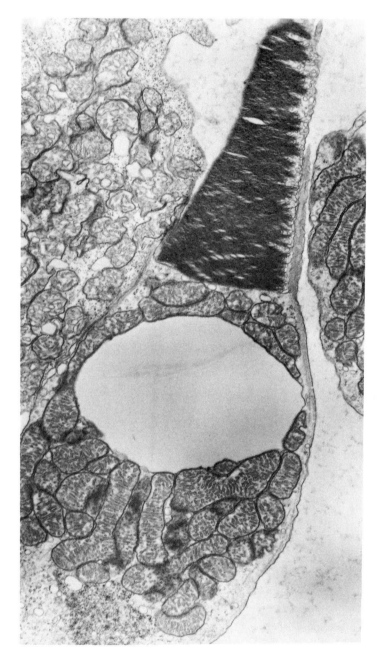

Figure 3.6

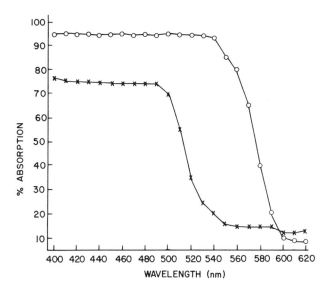

Figure 3.7 Absorbance curves for two different classes of colored oil droplets. The droplets differ in the spectral locations where they begin to transmit appreciable amounts of light. In this example, one of these droplets would appear red (O—O) while the other (×—×) would appear a greenish-yellow.

vices for reducing the effects of chromatic aberration imposed by the limitations of the optics of the eye.

C. Other Ocular Filters and Reflectors

There are several other classes of ocular structures potentially capable of imposing spectrally selective effects. First, it should be noted that the corneas of many species of interest from a color vision viewpoint are uniformly transparent across the visible spectrum. However, in various teleost fish the cornea may be substantially yellowed, as are many lenses (Walls, 1942). Unlike lens pigmentation, however, the coloring of the teleost cornea is usually not spatially uniform (Moreland and Lythgoe, 1968). In many of these species the pigment is considerably denser in the dorsal portion of the cornea than elsewhere, and con-

Figure 3.6 Photomicrograph showing a cone from the retina of the African clawed frog (*Xenopus laevis*). The large inclusion located among the mitochondria in the ellipsoid is an oil droplet. Although the droplet in this cone is colorless, most such droplets contain carotenoid pigments which give them a vivid color. Note that much of the light passing axially along the photoreceptor will go through the droplet before it reaches the outer segment. (Micrograph supplied by M. S. Eckmiller.)

sequently the effects of spectral filtering by these corneas depends on the directionality of the light source relative to eye position, as well as its spectral distribution. It has been speculated that this differential pigmentation of the cornea in teleosts serves as a device to attenuate retinal illumination from above, to perhaps compensate for the fact that considerably more intense light reaches the fish from that direction (Muntz, 1972).

Another ocular filter of importance in man is the macular pigment. This pigment is contained within the retina itself, in a circular patch of 6° to 10° centered on the fovea. It is present in both the inner and outer plexiform layers, but is especially dense in the latter location (Snodderly, Auran & Delori, 1979). The pigment is apparently the xanthophyl, lutein. Like the other ocular filters listed, the effectiveness of this pigment is restricted to the short wavelengths. However, as Figure 3.8 shows, unlike lens and corneal absorption, the macular pigment does not absorb strongly in the ultraviolet portion of the spectrum. It is conventionally said that this macular pigment is present in both man and the "other diurnal primates" (Muntz, 1972). Despite this claim, published measurements of the absorbance by macular pigment in species other than man appear to be scarce.

All of the ocular filters so far mentioned (cornea, lens, and macular pigment) have in common that they decrease the effectiveness of short-wavelength radiation. In addition to these filters, one other class of spectrally selective ocular devices bears brief mention. For most species, light that passes between the photoreceptors or through the photoreceptor avoiding absorption is then absorbed by the pigment epithelium. The epithelial pigment is melanin; it is spectrally nonselective. However, in a number of vertebrate species, the retina contains a

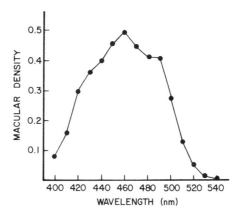

Figure 3.8 Spectral absorbance by the human macular pigment. (Data taken from Wyzecki & Stiles, 1967.)

reflective tapetal layer instead of a pigment epithelium. A wide variety of structural forms of tapeta are found in different species (Rodieck, 1973). Since tapeta are generally found in predominantly nocturnal species they are obviously of minimal significance as a color vision device. However, tapetal reflection is spectrally selective in some species known to have some color vision (such as the domestic cat), and hence might exert some small influence on color vision in those species.

III. Photoreceptors

As reasonably accessible receptors in the most important sensory system of many vertebrates, it is hardly surprising that photoreceptors have been the subjects of intensive study. From the viewpoint of color vision, interest in photoreceptors centers on the question of trying to identify those photoreceptors capable of providing the signals used for color vision, and on detailing the characteristics of these signals. This section considers some issues involved in identifying which receptors are rods and which are cones.

A. Duplicity Theory

The identification problem has historically been a complex one whose roots are to be found in a broad generalization known as duplicity theory. The starting point for duplicity theory was the observation made in 1866 by the anatomist Schultze to the effect that the photoreceptors in vertebrate retinas could be divided into two classes—rods and cones. He further observed that the retinas of nocturnal animals contained predominantly rods whereas those of diurnal animals had large numbers of cones. Because nocturnal and diurnal animals differed significantly in their visual behaviors, this suggested to him that the two types of receptors subserve quite different roles. Among these differentiated capacities was color vision, a feature clearly associated with the presence of cones. The degree to which duplicity theory was made explicit by Schultze and by those who followed him, and exactly what the tenets of this theory were, became a subject of considerable uncertainty and argument, issues which have been thoroughly reviewed elsewhere (Saugstad & Saugstad, 1959). However, there is no question that this generalization attributes certain functional properties of vision to the two classes of photoreceptors per se and not to any accessory structures or to subsequent processing by the visual system. Furthermore, although the duplicity theory merely associated cones with color vision, there is also no question that in the minds of many this association became a strongly predictive link—the presence of cones indicates color vision, their absence its lack. Under this view all that needed to be done was to determine if the retina contained cones. If it did, the animal had color vision, if not, color vision was absent. Assuming that an animal possesses a complex functional capacity like color vision solely by virtue of its possession of cones seems an obviously unwarranted jump, somewhat akin to assuming that the presence of five toes implies a capacity to kick field goals. However, even if one were prepared to accept a compulsive link between photoreceptor type and functional capacity, there remains the very serious problem of determining which photoreceptors are rods and which are cones.

Figure 3.9 shows sketches of the type frequently used to illustrate the major morphological features of vertebrate rods and cones. The same general structural components can be distinguished in both classes of photoreceptors. Thus, both have outer segments that project into the space between the retina and the pigment epithelium, an inner segment that contains a number of types of organelles (and which is connected to the outer segment by a ciliary process), and terminal regions where synapses are formed between the photoreceptors and the second-order retinal cells. The classical distinction made between rods and cones was based on the observation that the shape of the rod is cylindrical so that the inner and outer segments have roughly the same diameter, whereas in the cone the outer segment tapers and has an inner segment of greater diameter than the outer segment (Cohen, 1972). Schultze used this morphological distinction along with several others (for example, location of the apical tips of the outer segments, the

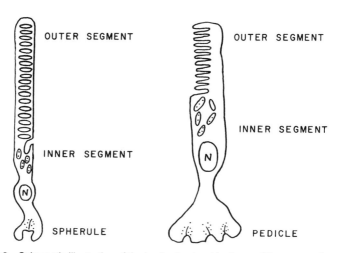

Figure 3.9   Schematic illustration of the basic structural features of the mammalian rod (left) and cone (right). The outer segments of each contain membranous discs; they differ in that the rod discs are isolated from each other and from the outer segment membrane whereas the cone discs are not (see Figure 3.12). The presynaptic specializations seen in the terminal regions (spherule and pedicle) include electron-dense ribbons and surrounding synaptic vesicles. N is the nucleus.

presence of oil droplets in some cones but never in rods, and the location of the receptor nuclei) to distinguish rods from cones (Crescitelli, 1972).

Figure 3.10 shows drawings made to represent the morphologies of photo-receptors found in the retinas of a variety of different classes of vertebrates. It is obvious that vertebrate photoreceptors vary enormously in structural configuration—in shape, size, and prominence of their various constituent parts. Small wonder, therefore, that attempts to relegate all of these cells to one of two classes run into difficulties. Compounding this difficulty of identification is the fact that not only is there great variation between species, there are often very substantial differences in receptor morphology within a single retina. The most familiar example of this latter variation is in the human retina where those cones found in the fovea are slender and basically nontapering, the classic prime structural characteristics of the typical rod. All of these structural variations, both within and between species, have raised questions about the rationality of classifying all photoreceptors as either rods or cones, a difficulty that Schultze himself apparently experienced (Crescitelli, 1972).

B. Criteria for Distinguishing between Rods and Cones

The difficulties of categorizing vertebrate photoreceptors often led inves-tigators to conclude that it was impossible to put all receptors into two categories without being greatly arbitrary. For example, on the basis of an analysis carried out with modern research methodologies, Pedler (1965) has argued that on morphological grounds alone, three categories are required to account for the full range of vertebrate photoreceptors. Perhaps more typically, investigators have attempted to use additional photoreceptor characteristics beyond their structure to distinguish rods from cones—things such as the type of photopigment contained in the receptor and the nature of the electrical signals generated by the receptors. How these functional properties relate to the rod–cone distinction will be sum-marized shortly. Before doing so, it will be useful to sketch the structural charac-teristics of photoreceptors that are currently used to distinguish rods from cones. One should be cautioned that not all of these distinctions are clear-cut, present in all species, or necessarily agreed on by all authorities. Rather, they appear to be the most reasonable current criteria used to distinguish vertebrate rods from cones on morphological grounds alone.

1. Receptor Shape

As already noted, the classic shape distinction between rods and cones was based on the appearance of the outer segment alone. Now, however, the shape difference between rods and cones is usually taken to include the entire cell so that the cone may appear tapered solely by virtue of the fact that the inner segment has a larger diameter than does the outer segment (Cohen, 1972). Foveal

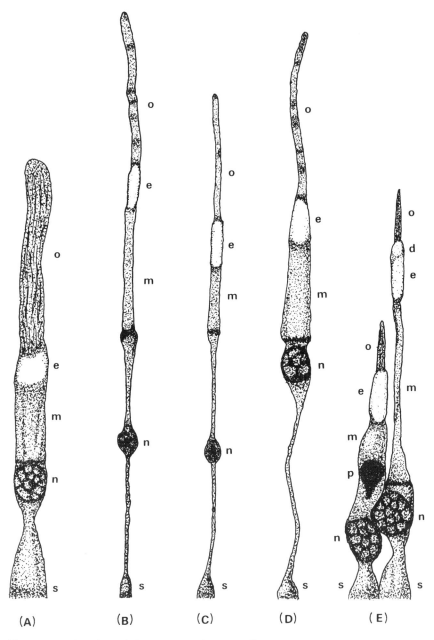

Figure 3.10 Drawings of photoreceptors found in different species illustrating the structural variations in vertebrate photoreceptor construction. (A) Frog rod; (B) human rod; (C) rat rod; (D) human cone; (E) a double cone from the frog with the accessory member of the cone to the left. Abbreviations: o, outer segment; d, oil droplet; e, ellipsoid; m, myoid; p, parabaloid; n, nucleus; s, synaptic terminal. (Redrawn from Young, 1969.)

cones are a frequent exception to this rule. It is also often the case that the relative positioning of the receptor organelles differ—for example, in retinas where both types exist, the rod nuclei may be displaced toward the vitreous relative to the cone nuclei, or vice versa. This arrangement has been called "tiering." It is thought that this adaptation permits the outer segments of both receptor types to simultaneously occupy the same light-capturing areas (Miller and Snyder, 1977). An example of such an arrangement is shown in Figure 3.11.

2. Outer-Segment Structure

The structure of the photoreceptor outer segments provides perhaps the best single criterion for separating rods from cones. As seen in the electron micro-

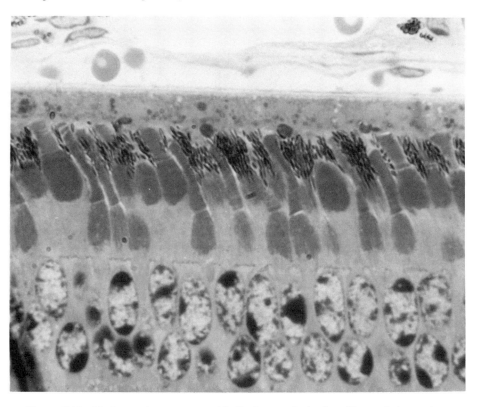

Figure 3.11 Photoreceptor arrangement in the tree squirrel retina. Those photoreceptors whose inner segments are located in the row toward the bottom of the picture are rods. The outer row of photoreceptors are cones. Note that because of this layering the bases of the outer segments of the rods are closer to the incoming light than are those of the cones and, consequently, light not absorbed in a rod outer segment may still be absorbed in a cone outer segment. Magnification: × 1325. (Photomicrograph supplied by G. Tabor.)

scope, photoreceptor outer segments appear as a stack of membrane-limited discs (see Figure 3.12). These discs have been the focus of particular interest because they house the photopigment molecules. Observations made on a large range of species indicate that there is a fundamental difference in disc arrangement between most vertebrate rods and cones (Cohen, 1972). As shown in Figure 3.12, the discs in the rods are isolated from one another and from the cell membrane. On the other hand, cone discs are typically continuous with the outer cell membrane, although in mammalian cones this continuity may be apparent only in the basal portion of the outer segment. Because of the difference in disc construction, the cone outer segment (at least over some of its length) is open to extracellular space whereas the rod outer segment is not. This property can be exploited as a means of differentiating rods and cones in that various substances injected into the extracellular space may permeate into the cone but not into the rod. If the

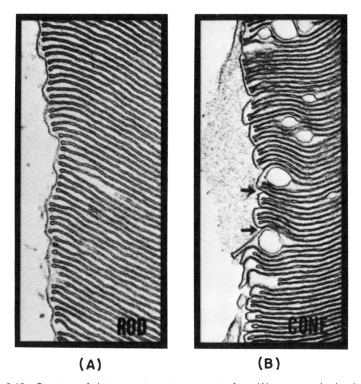

(A)   **(B)**

Figure 3.12   Structure of photoreceptor outer segments from (A) a gray squirrel rod and (B) ground squirrel cone. Note that the rod discs are not continuous with the outer segment membrane whereas the cone outer segment has numerous locations where the membrane and the discs are continuous (arrows). Magnification: × 38,000. (Photomicrograph supplied by D. H. Anderson.)

injected substance is such that it can be visualized, for example, dyes like procion yellow, then this technique becomes a potentially useful means for distinguishing cones from rods (Laties and Liebman, 1970).

3. Outer-Segment Renewal

As highly specialized cells, photoreceptors are continuously involved in the replacement of their molecular constituents. Much attention has been directed to the dynamics of this replacement process in the outer segments. Of interest here is the discovery made in the mid-1960s that the process of renewal in vertebrate outer segments appears to be different for rods and cones (for a comprehensive review of this area see Young, 1976). If radioactively labeled amino acids are supplied to the photoreceptor, they are soon incorporated into new proteins and their locations and movements can be traced and visualized by autoradiography. Rod outer segments show a picture of membrane renewal in which new discs are constantly being formed at the base of the outer segments. These newly formed discs appear in the autoradiogram as a dense, compact band of radioactivity. Over time these newly formed discs move up the outer segments, eventually detach from the photoreceptor, and are phagocytized by the pigment epithelium. On the other hand, the renewal process in the cone outer segment was originally believed to involve replacement of the disc membrane *in situ* so that radioactivity is seen to be randomly scattered throughout the outer segment. Thus, it was believed (Young, 1976), no new discs are formed nor are any old ones destroyed in the adult cone. Since these differences in the mode of membrane renewal can be clearly seen in the autoradiogram, Young suggested that this method might constitute another means of distinguishing rods from cones.

Although the method of outer segment renewal is well accepted for vertebrate rods, there has been less concordance of opinion about the details of the process in the cones. Recent observations indicate clearly that in cones, as well as in rods, new discs are constantly being formed and then later phagocytized (Anderson, Fisher, & Steinberg, 1978). Despite the fact that complete agreement has not yet been reached as to if, or how, rod and cone outer segment renewal differs, there do not yet appear to be any instances where autoradiographic labeling experiments give the same results for photoreceptors known on other grounds to be rods and cones. Consequently, the labeling procedure may provide another means for structurally differentiating cones from rods.

4. Staining Differences

During the long history of efforts to identify rods and cones, there are scattered reports claiming that the two types of receptors can be distinguished by differences in their appearance in stained sections. For example, with agents such as Mallory's triple stain, it is possible to differentially stain rods and cones in a number of different species, including man, so that in stained sections the rods are colored blue while the cones take on a bright reddish appearance (Last, 1961). In other cases the degree of staining has been used as a morphological criterion for distinguishing rods from cones. For instance, West and Dowling (1975) found that in the ground-dwelling sciurids the inner segments of rods stain considerably less densely than do corresponding portions of the cones. Finally, for at least some species, the appearance of the nucleus in stained tissue can be used as a criterion for differentiating receptor types—in rat, the nuclear heterochromatin of the rods appears as a single large clump while in the cones multiple small clumps of heterochromatin are characteristically seen (Sokol, 1970; LaVail, 1976).

5. Receptor Terminals

From the time of the earliest observations of vertebrate photoreceptors with the light microscope, it has been known that the receptor terminals for rods and cones differ in appearance. And naturally, with the more recent application of the electron microscope to the study of photoreceptor structure, much more has been learned about terminal structure. The classic difference between rod and cone terminals is in their relative size and shape: rod terminals tend to be smaller and rounder (spherules) whereas cone terminals tend to be larger and flatter (pedicles). The major regions of synaptic contact in the receptor terminals have a particularly striking appearance in the electron microscope (see Figure 3.13). The postsynaptic elements, the bipolar and horizontal cells, are often inserted into a deep invagination in the photoreceptor terminal. Typically, there are three or more such postsynaptic elements. The presynaptic specializations include an electron-dense ribbon (Figure 3.13) surrounded by a cloud of synaptic vesicles. In most vertebrate retinas the rod terminals have only a single such invagination whereas the cones tend to have multiple invaginations and hence multiple locations of synaptic contact. Vertebrate photoreceptors also have a range of other apparent synaptic specializations, and some of these appear to be specific to cones (see Dowling, 1974).

The preceding suggests that terminal structure offers a concise package of anatomical specializations that allow one to simply differentiate rods from cones. Unfortunately, this is not completely so. Although the terminal distinctions just mentioned are often clear and consistent, there are frequently puzzling and substantial variations across species. To note just one example, Cohen (1972) reports that by their appearance all of the photoreceptor terminals in the pigeon qualify as pedicles despite the fact that the pigeon retina is known on other grounds (see Chapter 4) to contain both rods and cones.

If one wishes to use morphological criteria to distinguish rods from cones, it is clear that not any of the structural differences summarized in the preceding are

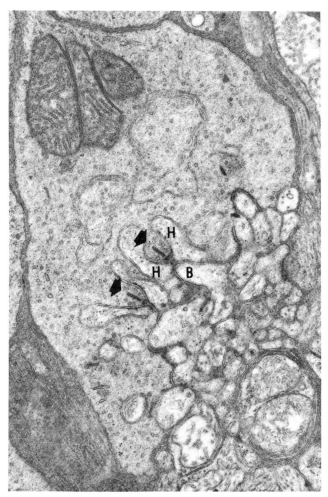

Figure 3.13 Photomicrograph of a cone pedicle in the cat retina. The solid arrows point towards two synaptic regions. The presynaptic specializations include the densely stained synaptic ribbons and numerous synaptic vesicles. In one of these zones the component cells in the postsynaptic triad are identified (H, horizontal cell; B, bipolar cell). Magnification: × 22,500. (Photomicrograph courtesy of S. K. Fisher.)

unambiguous for all photoreceptors, nor do they necessarily always agree with one another. Nevertheless, form can be used to categorize most photoreceptors as rods or cones. In cases where this categorization is not possible on structural criteria alone, certain functional properties of these cells may be used as adjunctive indices. If this is done, most, if not all, vertebrate photoreceptors can be seen to fall into two classes. Going beyond that, however, requires considerable caution. In an excellent review article, Cohen (1972) has clearly stated the sensible view that although the presence of morphologically defined rods and cones does *predict* certain visual functions, it is misleading in the extreme to think of these functional properties as residing in the photoreceptors. The frequent practice of referring to cones as "color receptors" is a particularly germane example of this error.

IV. Photopigments

Photons that survive the perilous passage through the eye and into the photoreceptor outer segments may initiate neural activity that results in vision. It has been known for over a century that retinas contain photolabile pigments that undergo rapid changes when exposed to light and that this change constitutes the primary active step in the visual process. Much has been learned about the retinal photopigments since the 1930s, particularly about the details of their chemistry, and consequently the literature is now very large. Our purpose here, as in previous sections, is restricted—in this case to briefly describe how photopigments are characterized, and then to note the variations in photopigments across species. Both of these aspects are useful prerequisites for an understanding of the comparative color vision literature.

A. Characterization of Photopigments

Photopigment molecules consist of an aldehyde of vitamin A (the chromophore) linked to a protein component (an opsin). The chromophores of all vertebrate photopigments are based on either vitamin A_1 or vitamin A_2. The former is now called retinal, the latter 3-dehydroretinal. These chromophores can combine with various opsins to form photopigment molecules. Those opsins that combine with retinal produce photopigments that as a class are called rhodopsins, those that combine with 3-dehydroretinal produce a class of photopigments called porphyropsins. Absorption of photon energy may be sufficient to cause a conformational change in the chromophore. This change, bleaching, results in the eventual hyperpolarization of the photoreceptor membrane. The steps between absorption and changes in membrane permeability are currently under intensive investigation, but as yet are not completely understood. The bleaching of a photopigment molecule renders it temporarily insensitive to light, and thus retinal illumination diminishes the amount of photopigment available to subserve vision. The amount of unbleached pigment in a light-adapted eye is proportional to the intensity of the light to which it has been exposed, although the nature of this proportionality may also be dependent on the time course over which illumination is delivered.

Photopigments are usefully characterized by measurements of their spectral absorbance. An example of an absorption spectrum for a prototypical vertebrate visual pigment is shown in Figure 3.14. The main absorbance peak of this photopigment is at 500 nm. A second peak, which is substantially lower in sensitivity, is located in the short wavelengths, centered broadly at about 350 nm. These secondary peaks, sometimes called β-peaks, are at wavelength values approximately two-thirds of those of the main peak. Reference has already been made to the speculation that one role of short-wavelength filtering by the lens might be to further reduce the effectiveness of these secondary absorbance bands.

Defining the absorbance spectra for the relevant photopigments has traditionally been an important aspect of the search for color vision mechanisms. Accordingly, it is worth noting briefly the methods used to accomplish this task. Detailed descriptions and critiques of these methods have appeared in several chapters of a recent volume on visual photopigments (Dartnall, 1972). At least in theory, the most straightforward method is to extract the photopigment from the retina and then examine it spectrophotometrically. Much of what is known about

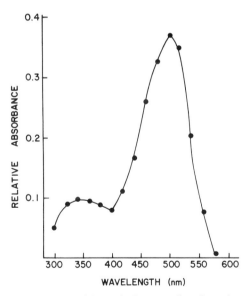

Figure 3.14 Absorbance spectrum for a typical mammalian photopigment. For this pigment the main absorbance peak is at 500 nm. A secondary absorbance peak, the β peak, is broadly centered at about 350 nm.

the kinetics of photopigment operation has been learned using this approach. The procedure has, however, several drawbacks:

1. The photopigment could be present in insufficient quantity to permit extraction.
2. Special procedures must be employed to be certain that only a single pigment is being measured in extract.
3. The absorbance properties of pigments in extract may differ somewhat from those of the same photopigments sequestered in outer segments.
4. Some photopigments, particularly those found in cones, are so far refractory to extraction procedures.

In the past 15 years technology has been developed to permit the measurement of the properties of photopigments contained within individual photoreceptors of excised retinas. This remarkable procedure, microspectrophotometry (MSP), although steadily improving, has in the past suffered from a lack of precision. In particular, MSP has not until recently been able to provide very reliable measurements of the spectra of photopigments contained in mammalian cones. Current MSP measurements are claimed to be able to place the wavelength of peak absorbance to an accuracy of ± 3 nm (Bowmaker, Dartnall, Lythgoe, & Mollon, 1978), and thus the future of this technique is most promising.

Another major approach to defining photopigment spectra is retinal densitometry. In this method light is beamed into an intact eye and the amount of light reflected back out is measured. Because retinal densitometry can be used in the intact eye, the results can be continuously correlated with other measures of visual capacity.

The major difficulty associated with the use of retinal densitometry, or any of a large range of other physiological and psychophysical procedures used to estimate the spectral absorbance properties of photopigments, is that retinas usually contain a number of different types of photopigments that show considerable spectral overlap and thus contribute jointly to the measurement. The strategy typically employed to attack this problem is to make the measurement in conjunction with intense chromatic adaptation, the idea being that with selective chromatic adaptation it is possible to bleach one photopigment more than another and so selectively reduce its effectiveness as an agent for absorbing light. The issues involved in making this assumption are complex. They have been discussed in detail elsewhere (Sirovich and Abramov, 1977). It suffices to say here that estimates of photopigment spectra obtained from these derivative measures should only be accepted cautiously.

In considering the role that photopigments play in determining vision, it is conventional to assume that each photon absorbed by a particular photopigment produces an identical effect (the principle of univariance). Thus, the effectiveness of a stimulus depends only on the number of molecules it causes to be bleached. The wavelength of the light is of importance only in the sense that the probability of absorption is higher at some wavelengths than at others. If this is so then a photopigment is a receptive device incapable of providing any information about the wavelength distribution of the stimulus. Thus, in the example of the photopigment spectrum shown in Figure 3.14, the effect of a 350-nm light on the photopigment can be made indistinguishable from the effect of a light of 500 nm on the same pigment if the two lights are appropriately adjusted in relative intensity.

Actually, there is at least one possible way in which it has been suggested that the effect of light on a photopigment might depend on its wavelength and not just on its intensity (Rodieck, 1973). If the nature of the photoreceptor signal (for example, its latency) generated by the photopigment differs at various locations along the outer segment, then a stimulus wavelength near the peak of the absorbance spectrum of the resident photopigment might be more likely to bleach pigment at the base of the outer segment than at other locations. Stimulus wavelengths away from this spectral location would not show this same differentiation. There does not appear to be sufficient evidence to yet reject this possibility (Rodieck, 1973), although it seems quite unlikely.

B. Species Variations

Photopigments have been studied in a large number of vertebrate species, particularly among fish. Those found in all mammals, reptiles, birds, and most marine fish are rhodopsins and are based on retinal; whereas photopigments found among freshwater fish are porphyropsins and are based on 3-dehydroretinals. The retinas of some species contain a mixture of the two classes of photopigments. Some migratory species show changes from a porphyropsin to a rhodopsin system during the move from fresh water to sea water while some amphibians show an analogous change during metamorphosis. An extensive review of the distributions as well as the ecological and evolutionary significances of rhodopsin versus porphyropsin photopigment systems has been provided by Bridges (1972). Table 3.1 provides a recent compilation of the distribution of rhodopsin and porphyropsin among a large number of vertebrate species.

During the early days of photopigment research, it was supposed that the absorption characteristics of photopigments were not greatly variant across species. Consequently, the names rhodopsin and porphyropsin referred not just to photopigments based on retinal and 3-dehydroretinal, respectively, but more specifically to photopigments that were based on these two chromophores and that had spectral peaks of absorbance (λ_{max}) at about 500 and 522 nm, respectively. As measurements were made of the photopigments of more species, it became apparent that λ_{max} values often differed from these locations. As a consequence, these new pigments were typically given new, descriptive names

Table 3.1

Distribution of Rhodopsin and Porphyropsin among Various Species of Vertebrates[a]

Group	Rhodopsins	Porphyropsins	Mixture
Freshwater fish	18	25	37
Marine fish	59	2	6
Migratory and estuarine fish	5	0	18
Amphibians (adult)	12	1	1
Reptiles	16	0	0
Birds	8	0	0
Mammals	22	0	0

[a] The data tabulated are the number of species within each of the categories. (From *The Vertebrate Retina: Principles of Structure and Function* by R. W. Rodieck. W. H. Freeman and Company. Copyright © 1973.)

(Dartnall, 1962). Not surprisingly, this procedure quickly lost its utility as the number of measured photopigments grew. A substantial simplification emerged in 1953 when Dartnall discovered that the absorbance spectra for all retinal-based photopigments had the same shapes if the spectra were plotted on a frequency (wave number) scale. In that case, the only variation among photopigments was in the value of λ_{max}. A nomogram was published (Dartnall, 1953) that permitted the generation of an absorbance spectrum for any retinal based photopigment given the λ_{max} value. The photopigments based on 3-dehydroretinal were found to have somewhat broader absorbance spectra, but as a group they too showed absorbance spectra of a common shape. Consequently, a second nomogram for 3-dehydroretinal visual pigments was also provided (Bridges, 1967; Munz and Schwanzara, 1967). With these simplifications, then, a visual pigment could be specified by two pieces of information: whether the pigment was based on retinal or 3-dehydroretinal, and the value of λ_{max}. A comprehensive listing giving these two types of information for about 340 vertebrate species is available (Lythgoe, 1972).

Unfortunately, the situation is not quite as straightforward as the previous paragraph would suggest. Within the context of deriving estimates for the visual pigments underlying human color vision, a number of studies carried out during the early 1970s indicated that the shape of the absorbance spectra for the various photopigments may not, as the derivation of the nomograms implies, all be the same irrespective of the λ_{max} value. Recently, Ebrey and Honig (1977) have argued that the absorbance bandwidth of a photopigment varies continuously as a function of its λ_{max} value. The direction of this variation is such that those photopigments having shorter λ_{max} values have broader absorbance spectra. This being the case, it is apparent that no single nomogram will suffice for all photopigments. Nevertheless, the change in absorbance bandwidth with change in the value of λ_{max} is gradual, so Ebrey and Honig (1977) suggest that three nomograms will provide reasonable accuracy for all photopigment absorbance spectra. The nomograms they propose for A_1 and A_2 pigments are given in Figures 3.15 and 3.16. Although possibly not the last word, these nomograms represent the best current information.

One final but significant point about variations in absorbance spectra: there is good evidence that there are measurable variations among individuals of the same species in the value of λ_{max} of a particular photopigment (Alpern & Pugh, 1977; Bowmaker, Loew, & Liebman, 1975). For example, Bowmaker *et al.,* (1975) used MSP to measure the absorbance characteristics of the photopigment found in frog rods. Among a number of individuals the value of λ_{max} so measured covered a range of 8 nm. These investigators argued that this variation is not an artifact of the measurement situation, but rather probably reflects genetically based individual variation. One should, therefore, consider a reported value of λ_{max} to represent the most preferred spectral location for the photopigment in that species. These individual variations in photopigment spectra may be important sources of within-species variations in color vision, as well as providing a matrix within which evolutionary changes in photopigment spectra might be achieved.

C. Rod and Cone Photopigments

Thus far in describing the variations among the photopigments no distinction has been drawn as to which photopigments were from rods and which were from cones. This issue is important in that if the photopigments contained in rods and cones clearly differ in some regard, then that difference might be used to provide another criterion for distinguishing rods from cones. The photopigments found in vertebrate rods and cones cannot be differentiated by their spectral absorbance properties—the nomograms given in Figures 3.15 and 3.16 are equally valid for, are indeed based on, both rod and cone photopigments. There do not seem to be any values of λ_{max} that are unique to either rod or cone photopigments. Neither are there any known differences in chemical make-up between the photopigments found in rods and cones—both are composed of one of the same two chromophores, and the particular opsins used are not unique to either rod or cone photopigments.

Of course, there is one well-known difference between rod and cone photopigments, namely, that the kinetics for bleaching and regeneration differ markedly for rod and cone photopigments. In the human, for example, the time constant for photopigment regeneration is about three times as long for rods as for cones. However, these differences are based on pigment changes measured *in situ,* and thus it is possible that the differences in kinetics are conferred by some

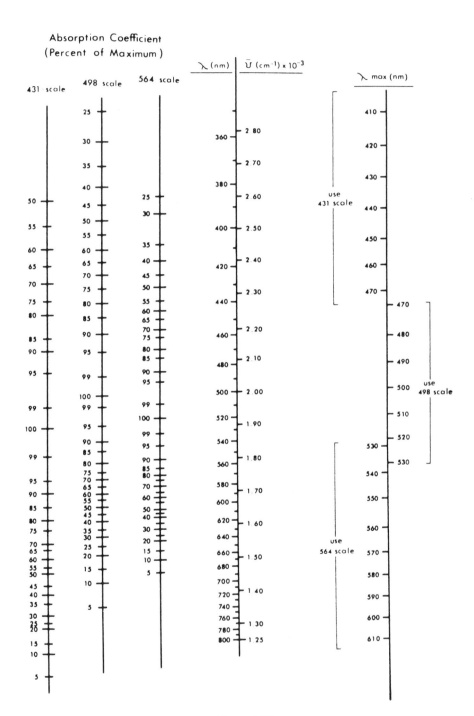

Figure 3.15

property of the photoreceptor organization, for example, the differences in outer-segment disc structure, rather than by any feature of the photopigment itself. This uncertainty could be resolved if it were easy to extract cone pigments and then to compare their behavior to that shown by the same pigment when housed within the outer segment.[1] Unfortunately, thus far almost all cone pigments are refractory to extraction procedures. The exception to this is the photopigment contained in chicken cones, called iodopsin, which in extract has been found to regenerate much faster than does the rod pigment from the same retina (Wald, Brown, & Smith, 1955). This is sparse evidence on which to base a generalization and so it seems clear that not enough is known from a comparative point of view to permit the identification of a photoreceptor solely on the grounds of the kinetics of the photopigment contained therein, although if those facts agree with other criteria then they certainly strengthen the argument for identification.

D. Retinal Distribution of Photopigments

There are few, if any, species in which the relative numbers of rods and cones remain constant throughout the retina. The extent of this heterogeneity can usually be established from a histological examination of the retina. A more difficult problem is provided by the fact that multiple classes of photopigments found within a single receptor type are also, usually, not uniformly distributed. From the many ways in which color vision in man is known to vary as a function of the retinal location of stimulation, it has long been known that the unequal distribution of photopigments has important functional consequences. Although in some species it may be possible to determine the topographical distribution of photopigments by virtue of structural differences in the photoreceptors that contain them, as, for example, in the goldfish (Stell and Harosi, 1976), this is usually not the case, and consequently it has been difficult to determine the retinal distribution of the various photopigment classes.

One recent solution to this problem is to make use of a histochemical marking procedure. Enoch (1964) demonstrated that the reduction of nitroblue tetrazolium chloride is light dependent and photopigment mediated. This reduction causes

---

[1] Some species, for example, frog and turtle (see Chapter 4), have photoreceptors identified as rods and cones that appear to contain the same type of photopigment. A comparative study of the dynamic properties of these photoreceptors might also shed some light on this question.

Figure 3.15 Wavelength-dependent nomograms for vitamin A_1 visual pigments. Three different wavelength scales are used (right columns). To compute the absorbance for any photopigment, place a straightedge on the wavelength corresponding to peak absorption (right column). The absorption coefficient (left columns) can then be read off for any wavelength value (center column). (From Ebrey & Honig, 1977).

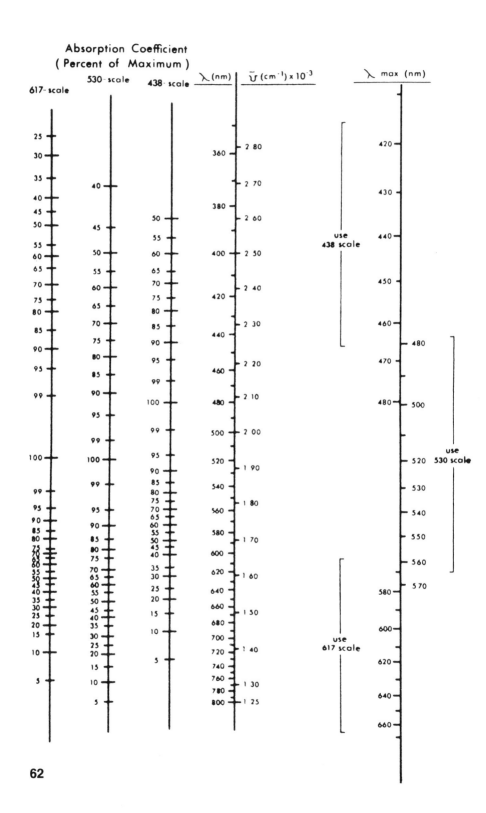

Absorption Coefficient
(Percent of Maximum)

the photoreceptor cytoplasm to take on a blue appearance. Marc and Sperling (1977) followed this reduction after eyes had been exposed to different spectral lights, the notion being that the photoreceptors should be differentially stained according to the wavelength of the stimulating light and the spectral characteristics of the photopigments. One of the species they examined, the baboon, has three classes of cone photopigments with λ_{max} at about 440, 536, and 565 nm. In the center of the fovea of this primate, the photoreceptors containing these three types of photopigment have a relative prevalence of 4%, 63%, and 33%, respectiyely, whereas at 5° from the fovea these proportions change to 13%, 54%, and 33%. The magnitude of this variation effect is illustrated in Figure 3.17. The histochemical marking procedure technique is a promising one that might be profitably employed to gain information about photopigment distribution in the retinas of many other species.

V. Neural Mechanisms

The electrophysiological approach to the study of sensory systems typically involves an examination of neural outputs at successive anatomical levels when controlled inputs are applied to the system. In the case of color vision, the goal is to identify and functionally characterize those cells which carry information that might be used for color vision purposes, and to discriminate these from any cells that are not involved in the production of color vision. Given the defining constraints of color vision, this means that particular attention should be directed to those elements whose outputs remain discriminably different when the system is stimulated with various equiluminant spectral stimuli.

It may be enlightening to start with a standard explanation of how a neural element could achieve a consistent difference in output to different wavelengths of stimulation even when these wavelengths are widely varied in relative luminance. It is schematized in Figure 3.18, and goes as follows. If each class of photopigment behaves univariantly, then any element receiving inputs only from receptors containing the same pigment (see Figure 3.18A) cannot transmit usable color vision information because the responses to different wavelengths can be made equivalent by simply varying their relative luminances. Similarly, any arrangement in which there are two or more classes of photopigments providing inputs to a cell, but whose inputs are simply summed by the cell, is also incapable of signaling information that satisfies the color vision requirement (see Figure 3.18B). However, a cell receiving inputs from more than one photopigment class in such a way that the inputs cause qualitatively different changes in the cell will

---

Figure 3.16 Wavelength-dependent nomogram for vitamin A₂ visual pigments. See Figure 3.15 for instructions on the use of this nomogram. (From Ebrey & Honig, 1977.)

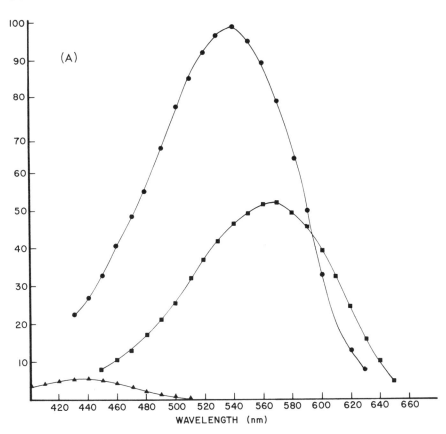

Figure 3.17 Absorbance curves for the three cone pigments found in the retina of the baboon (λ_{max} = 565, 536, and 44 nm). The relative heights of the absorbance curves have

satisfy the color vision requirement. One such scheme is shown in Figure 3.18C where the recipient cell is excited by inputs from one photopigment class, and is inhibited by inputs from the other photopigment class. In this case no adjustment in the relative luminances of the two stimuli will produce equal effects in the cell. More complicated schemes can be imagined to lead to the same outcome, and obviously further properties may be required to satisfy the details of any particular type of color vision (for example, the presence of more than two classes of photopigments), but Figure 3.18C defines a minimal situation.

The previous paragraph describes some neural machinery that might be used to extract the information required for color vision. It is important to reemphasize, however, that the presence of color vision *can only* be established by appropriate behavioral tests of the sort described in Chapter 2. Despite that important fact,

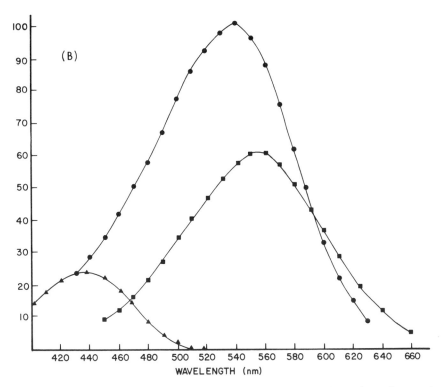

been scaled to reflect the prevalence of photoreceptors containing these three classes of pigments in (A) the fovea, and (B) in a region 5° from the fovea. Note that the relative proportion of the 565-nm cones is about the same in the two locations while the relative proportions of the 536- and 440-nm cones change significantly. (Data taken from Marc & Sperling, 1977.)

which is sometimes lost in discussions of the physiology of color vision, there are several sorts of physiological observations that can be used to infer that at least the possibility for some color vision exists. In order to derive such an inference, it needs to be established that

1. two or more mechanisms having different spectral sensitivities are present in the visual system;
2. that these mechanisms are simultaneously operative under some set of photic conditions; and
3. that the outputs from these different mechanisms are brought together at some point in the system in such a way that their differences in response to various spectral stimuli are preserved.

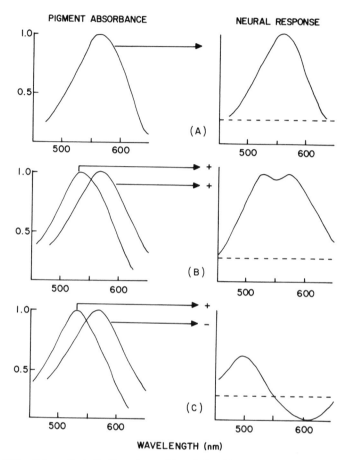

Figure 3.18 Schematic indication of possible relationships between photopigment absorbance properties (left column) and the responses of neural elements receiving inputs from these photopigments (right column). (A) The response of an element receiving input from a single photopigment class mirrors the spectral properties of the pigment. (B) The outputs from two different photopigment classes are summed to produce a neural response whose spectral properties are roughly the envelope of the two pigment curves. (C) The outputs from the two photopigment classes generate neural changes of opposite sign that are combined algebraically. The horizontal dashed lines represent the level of maintained activity in the neural elements.

As we shall see, each of these classes of observations has been repeatedly used to support the conclusion that the visual system under examination may likely subserve color vision.

The search for the neural mechanisms underlying color vision has been a reasonably active one and, consequently, quite a substantial literature has been generated. Fortunately, this area has also been systematically reviewed on a number of occasions (Abramov, 1970; Daw, 1973; De Valois, 1973; De Valois and De Valois, 1975; De Valois and Jacobs, 1981). The ready availability of these reviews obviates the need for yet another comprehensive treatment. Rather, what follows is a selection of results intended to broadly characterize what has been accomplished by neurophysiologists interested in color vision.

A. Historical Perspective

Credit for the initiation of research on the neural mechanisms involved in color vision is rightfully given to Ragnar Granit, who in the late-1930s started a program of research aimed at "determining the colour sensitivity of single or small groups of retinal elements" (Granit, 1947, p. 91). Granit's main findings are well known. In recording from retinal ganglion cells in a number of different species, he distinguished two functional classes of cells according to their spectral responsivity: some were broadly responsive to a wide range of spectral wavelengths (dominators), while others (modulators) only responded to a much more restricted band of wavelengths. The spectral response properties of these two types of units are illustrated in Figure 3.19. In considering these results Granit proposed a theory in which the dominator elements were assumed to be responsible for providing brightness information whereas the modulator elements provided the "cues for discrimination of wave length" (Granit, 1955, p. 143).

Although Granit's dominator–modulator theory was subsequently criticized on

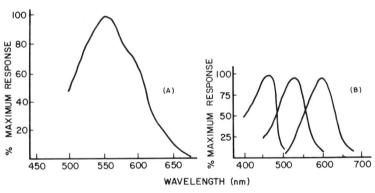

Figure 3.19 Spectral response curves determined for single retinal ganglion cells in the cat by Granit. (A) The broadly responsive element is a "dominator"; (B) the three more narrowly tuned cells are "modulators." For some species these modulator curves were obtained directly; for others (such as the cat) they were obtained by computation by subtracting the spectral response curve for a cell recorded in the presence of a chromatic adaptation light from the response of that same cell recorded in the dark-adapted eye.

a number of grounds, it did encourage others to attempt to analyze color vision through neurophysiology. Furthermore, his parcellation of cells into a class appropriate for providing color vision information and a class inappropriate for providing color vision information has been amply validated by all of the more recent investigations. What was eventually found to be inadequate (or at least incomplete) was the characterization of the response properties of those cells presumed to carry color information (the modulators).

In 1953, Svaetichin reported a striking finding that drastically altered the conception of the neural mechanisms for color vision. Svaetichin was attempting to record from the photoreceptors of estuarine fish. In doing so, he observed a variety of slow electrical potentials occurring in response to illumination of the retina (S-potentials). Among these, the most remarkable were those "elements" whose response polarities changed as a function of the wavelength of the stimulus. The responses of these so-called "C-potentials" were in the depolarizing direction (that is, the transmembrane voltage decreased) for some test wavelengths and in the hyperpolarizing direction for others (see Figure 3.20). Although the C-potential generators were later found not to be in the photoreceptors (see the following), as Svaetichin had supposed, it was quickly recognized that the nature of these responses was such that they were ideal candidates for the transmission of color information. Clearly, since response polarity was different for at least two wavelength bands, then the response to these different wavelength bands would necessarily stay discriminably different irrespective of

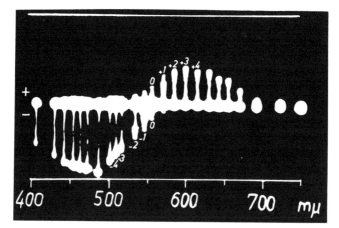

Figure 3.20 The spectral response pattern of an S-potential recorded from a horizontal cell in the fish retina. Each vertical trace shows the response to a single test wavelength. These graded potentials are either in the hyperpolarizing direction (upwards) or depolarizing (downwards) direction depending on the test wavelength. (From Svaetichin and MacNichol, 1958.)

their relative luminances. As already noted, this is a property required of a usable signal for color vision. In addition to this kind of response pattern (generically termed spectrally or chromatically opponent), Svaetichin reported finding other elements that were also responsive to a wide band of spectral wavelengths but which showed no reversals in the polarity of their responses (L-potentials). These elements, termed "spectrally nonopponent" or "broad-band," were functionally similar to Granit's dominator elements.

Not long after Svaetichin's initial reports on the retinal C-potentials, qualitatively similar response patterns were found to be characteristic of some cells in the lateral geniculate nucleus (LGN) of the macaque monkey (De Valois, Smith, Kitai, & Karoly, 1958). These responses were manifested as increases and decreases in action potential frequency, rather than in graded potential changes, but in other ways were highly similar to the retinal potentials. Shortly thereafter it was shown that the nature of the chromatically opponent responses recorded from a single cell could depend on the spatial configuration of the stimulus, as well as on its spectral content (Wagner, MacNichol, & Wolbarsht, 1960). This discovery tended to bring the research on color mechanisms into the general search for the organization of receptive fields in the visual system. Over the past 20 years, chromatically opponent responses have been found to occur in cells recorded in a variety of different visual system locations, from retina to cortex, and in a number of diverse species, from frog to man. Further discussion of the nature and interpretation of this important response class will be provided presently.

A very significant step toward closing the gap between photopigment kinetics and electrophysiology occurred in the mid-1960s. Although many previous workers had tried to record electrical activity directly from the photoreceptors, indeed some had claimed success at this venture, it remained for Tomita and his associates (Tomita, 1965; Tomita, Kaneko, Murakami, & Pautler, 1967) to accomplish the task. They succeeded by mounting an excised fish retina on a movable platform. The platform could be "jolted" over short distances with high acceleration toward the tip of the recording electrode. By jolting the retina it proved possible to penetrate into the photoreceptor with a fine microelectrode and to obtain intracellular recordings. As recorded in this way, fish cones, indeed all vertebrate photoreceptors, are found to respond to direct increases in illumination with graded potentials that are in the hyperpolarizing direction. The ability to record directly from the photoreceptors permits the study of the details of transduction and provides an opportunity to see how information arising in the cones is transformed at this very early stage of visual analysis.

Three points should be added to this perspective. First, it has been emphasized that the "responses" of a putative color mechanism must remain different in the face of wide variations in the relative luminances of different spectral stimuli. However, the question of how these responses are to be measured has not been raised. Because of the very abundant evidence for the use of frequency codes in the vertebrate visual system, almost without exception the response measures that have been used in the study of color mechanisms are the amplitude of a graded potential or its transform, the rate (or change in rate) of action potential generation following stimulation. Both are typically sampled over a relatively long period—often 200–1000 msec following the stimulus onset. Although there is strong evidence that frequency represents a response dimension used in the coding of color, there is no totally convincing evidence that other features of nerve impulses are not used. In particular, it is sometimes suggested that the temporal patterning of the impulse train might represent the metric for a color code. This possibility arises in the context of trying to account for the wide variety of changes in perceived hue that can be shown to depend on the temporal properties of the stimulus alone, as seen most notably in the Fechner–Benham colors. Although two studies have been reported that claimed to find evidence for such temporal codes in the visual system (Kozak & Reitboeck, 1974; Gur & Purple, 1979), this notion has not gained much acceptance. Despite the current lack of strong evidence for the utilization of other than frequency codes, the reader should be cautioned that it is not yet possible to conclude either for or against their actual existence.

Second, throughout this discussion of neural mechanisms, I shall refer to studies involving the analysis of single unit responses. It is appropriate, therefore, to include a comment on other possible levels of neural analysis; to note in particular that there is a very large body of research on visual neurophysiology that comes from studies using gross electrical responses. Two such gross potentials, the electroretinogram (ERG) and the visually evoked cortical potential (VECP), have been much exploited. Although these potentials have the virtues that they are easy to record and that they can be used to index many aspects of visual sensitivity, until now, at least, they do not appear to have yielded much information about the neural mechanisms for color. What information there is on the use of these potentials to study color vision can be found in two reviews: Riggs and Wooten (1972) for the ERG, and MacKay and Jeffreys (1973) for the VECP.

A final (and curious) fact should also be noted. Although it is obvious that discrimination of decreases in luminance must be as important for visually guided behavior as discrimination of luminance increases, with very few exceptions the stimulus–response relationships studied by electrophysiologists have involved luminance increments, not decrements. Vision scientists are apparently much more attuned to turning lights on than to turning them off! Visual cells often, although not invariably, show symmetrically reversed responses to luminance increments and decrements, that is, firing faster to an increase in luminance and firing slower to a decrease in luminance or vice versa. This point is worth remembering because the response descriptions provided and the terminology employed are usually appropriate to luminance increases only. As will be described later, a corresponding symmetry can be seen in the responses of neurons transmitting color information. Neurons of this type show increases in firing rate if the stimulus involves a wavelength shift in one direction and decreases in firing rate when the stimulus is wavelength-shifted in the other direction, even when there is no net change in luminance.

B. Photoreceptor Signals

As we have just noted, a very significant step forward in retinal physiology came about when it became possible for the first time to impale single photoreceptors and record their electrical signals. To the extent that photoreceptors act as independent elements whose electrical responses are proportional to the number of photopigment molecules bleached by any incoming light (the principle of univariance), then one might expect that (1) the spectral sensitivity of the photoreceptor electrical response should exactly match the absorption spectrum of the resident photopigment, and (2) that the response of the photoreceptor should be determined only by the quantal catch in that photoreceptor and not by the characteristics of photic events transpiring elsewhere on the retinal mosaic. Both of these propositions are of interest in attempting to determine where in the visual system neural processing directed toward eventual color vision is initiated.

By virtue of the technical difficulties and consequent uncertainties inherent in both types of measurements, it is not yet possible to make many comparisons of cone pigment spectra and cone photoreceptor spectral sensitivities. Tomita *et al.* (1967) measured the spectral response characteristics of a fairly large number of cones in the retina of the carp and found that their response curves could be divided into three classes having response peaks at 462, 529, and 611 nm. These peaks correspond fairly closely to the absorbance peaks (at 455, 539, and 625 nm) for the cones in the retina of a closely related species, the goldfish, as measured by MSP (Marks, 1965). A somewhat closer correspondence has been reported between the absorbance spectra for cone pigments in the retina of a freshwater turtle, *Pseudemys,* and the spectral response curves obtained from direct cone recordings (Baylor & Hodgkin, 1973). It is probably reasonable to assume that the spectral characteristics of the direct electrical responses of the vertebrate photoreceptors are well accounted for by the spectral absorbance characteristics of the photopigment located within the outer segments of these cells.

In line with our earlier discussion of duplicity theory, it is worth pausing here to describe the nature of the response recorded intracellularly from rods and cones. As noted before, both types of photoreceptors respond to increases in the intensity of a stimulus light by showing graded hyperpolarizations. At comparable stimulus intensities, the latencies and rise times of these responses do not appear to be different for rods and cones (Fain & Dowling, 1973). However, the rod and cone responses do differ drastically in their rate and course of recovery at the termination of the stimulus—the cone response returns to its prestimulus level much more rapidly than the rod response. This difference clearly implies a better temporal resolution in the cone system. The two can also be separated by the nature of their responses to steadily presented lights. Although both may show an initial large hyperpolarization at light onset, the rods tend to stay close to this same level as long as the light is on, whereas the hyperpolarization seen in the cone tends to show a greater decay during this same time period. This latter property has important implications for the course of adaptational changes seen in the two types of photoreceptors (Normann & Werblin, 1974).

Certainly one of the most important recent findings about photoreceptor physiology is that rather than acting as independent elements, the photoreceptors are in fact subject to a range of modifying influences. That is, the mixing of photopigment-generated responses occurs at the very earliest stages in the visual system. Over the past two decades the retina has been subjected to extensive electron microscopic examination. One result of this labor is that a variety of anatomical contacts between photoreceptors (interreceptor junctions) have been observed. The nature and appearances of these specializations vary substantially across species.

In 1971, Baylor, Fuortes, and O'Brien reported that the peripheral cones found in the turtle retina responded to light that did not fall directly on their outer segments. Two classes of such effects were described: synergistic and antagonistic. In the synergistic effect, a voltage change produced in one photoreceptor caused a smaller voltage change in the same direction in neighboring receptors. In the turtle retina this interaction operates over a distance of about five receptor diameters. Furthermore, this interaction is pigment specific in that only those cones having the same type of photopigment interact with one another (Baylor, 1974). In the case of the turtle retina, this photoreceptor interaction appears to be mediated by way of gap junctions located on the photoreceptor terminals. The functional consequences of this lateral interaction are in a direction that is counter to the advantages usually ascribed to a fine receptor mosaic, that is, it appears to enhance sensitivity and decrease spatial resolution.

A second, more complex, interaction occurring at the level of the photoreceptors has also been found (Baylor *et al.*, 1971). This interaction is an antagonistic one in that it involves a depolarizing effect on the photoreceptors (which, it will be recalled, normally respond to luminance increments by hyperpolarizing). In the turtle this interaction is mediated by a synaptic contact between the L-type horizontal cell and the photoreceptor. Although these interactions will be discussed further in the next section, it is worth pointing out here that since these horizontal cells receive inputs from cones containing more than one photopigment class, their antagonistic feedback can alter the spectral response characteristics of the photoreceptors on which this feedback is registered. Insofar as those species investigated to date in this manner are at all representative (and it is primarily the turtle that serves as the current model), then it is apparent that the interaction of cone signals that are steps toward color vision first occur at the very earliest stage in the visual system—at the photoreceptors themselves.

C. Early Retinal Interactions

The developing capacity to successfully record intracellularly from cells located within the outer plexiform layer and to parcel out the respective contributions of these cells to the neural processing of visual information constitutes one of the important new developments in visual physiology. Two summary conclusions can be drawn from this rapidly developing endeavor. First, in accord with the substantial range of variation in structure, there are correspondingly large intraspecies differences in the electrophysiology of the outer portions of the retina. Second, the range of interactions between the various cell types is much more extensive than earlier models of retinal operation suggested.

It is now generally agreed that the S-potentials discovered by Svaetichin are generated by the horizontal cells (Rodieck, 1973). L-type S-potentials are in the hyperpolarizing direction for all spectral wavelengths. In most, if not all, cases the L-type response represents inputs to the horizontal cell from more than one photopigment class, although exactly which photopigment classes are represented and the degree to which they contribute to the L-potential varies widely from species to species; indeed, it also depends substantially on the details of the particular stimulus employed, for example, its spatial structure (Fuortes, Schwartz, & Simon, 1973). More interesting from the point of view of potential color vision relevance are the C-type potentials, those that show hyperpolarizing changes for some stimulus wavelengths, depolarizing changes for others. One might imagine that this response could arise in the straightforward manner suggested by Figure 3.18 such that the input to the horizontal cell from a photoreceptor containing one type of photopigment produces hyperpolarization while the input from a photoreceptor containing another type of photpigment produces depolarization. However, it is clear that the interactions leading to the C-type response are generally much more complicated.

As an example of the kinds of interactions actually occurring in the outer plexiform layer, consider those seen in the goldfish retina. As previously noted, the goldfish retina contains three classes of cone photopigments. There are also three classes of horizontal cells that receive cone innervation—one generating an L-potential and two showing C-type potentials. On the basis of these electrophysiological descriptions and observations of the anatomy, Stell and Lightfoot (1975) have constructed a model to suggest how these response classes arise. Surveying those data they concluded that the horizontal cell responses cannot be accounted for entirely on the basis of monosynaptic interactions, that is, by direct cone-horizontal cell pathways. Rather, in each instance, in addition to a direct input the horizontal cell must be influenced by an interneuron. Surprisingly, the interneuron so identified is another horizontal cell. Stell and Lightfoot (1975) suggest that these horizontal cell/horizontal cell interactions are mediated by chemical synapses located at or near the region of the cone pedicle.

The mixing of photopigment signals at early stages in the retina can also occur by virtue of the fact that photoreceptors in some retinas are postsynaptic as well as presynaptic to horizontal cells (Fuortes *et al.*, 1973). Consequently, since the L-cells receive inputs from more than one photopigment class, the outputs from the photoreceptors postsynaptic to L-type cells can depend not only on the spectral distribution of the light falling on the photoreceptor, but also on the spectral distribution of light falling within the receptive fields of the horizontal cells providing input to the photoreceptor (see Figure 3.21).

In the turtle, as in the goldfish, the responses of the horizontal cells are also partially explained as being due to interactions between horizontal cells (Fuortes and Simon, 1974). These examples reinforce the contention that significant signal interactions occur very early in the system and that these interactions often seem species-unique.

What must quite certainly be color relevant signals have on occasion been recorded from other cells positioned early in the processing chain, particularly the bipolar and amacrine cells (Kaneko, 1973). However, thus far, these investigations have not added much to the description of color processing. Most of that story still comes from information on the response properties of third- and fourth-order cells in the visual system, the ganglion and lateral geniculate nucleus cells. These will be considered next.

D. Ganglion Cells and Lateral Geniculate Cells

As we have observed, those interactions between cone signals required to produce chromatically opponent response patterns occur very early in the visual system. Despite that, most of the information about these response patterns has been obtained from recordings made from third- and fourth-order cells. This comes about because these elements generate action potentials and consequently yield to analysis by means of extracellular recordings, a procedure that is technically straightforward.

Chromatically opponent responses have been recorded from third- and fourth-order cells in the visual systems of a fair number of different species. Indeed, there do not appear to be any instances in which such responses have not been detected in a species known on other grounds to possess some color vision, and there is some reason to believe that the relative proportions of cells showing chromatically opponent and nonopponent responses may be positively correlated with the degree to which color vision is a developed capacity in any species. The

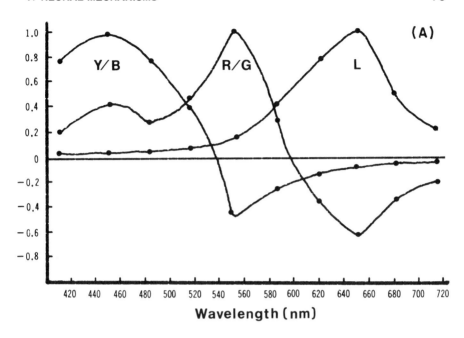

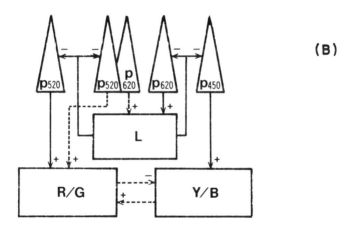

Figure 3.21   Schematic representation of horizontal cell responses in the turtle retina. (A) The spectral responses curves for two classes of C-cells (B/Y and R/G) and the L-cells. (B) The kinds of neural connections believed necessary to account for the responses of the cells in (A). The triangles show the different types of cones found in this retina. The pluses and minuses indicate transmission between elements with and without inversion of polarity. The connections responsible for the main response properties of each cell are shown by solid lines while the dashed lines represent secondary modifying interactions. (After Fuortes and Simon, 1974.)

evidence for this supposition has not been developed systematically, but it can be pointed out that in the macaque monkey, a species having normal trichromatic color vision, at least 70% of all fourth-order cells show chromatically opponent properties (De Valois, Abramov, & Jacobs, 1966; Wiesel & Hubel, 1966) whereas in the ground squirrel, a known dichromat, only about 30% of all cells show evidence of chromatic opponency (Michael, 1968; Jacobs & Tootell, 1981), and in the domestic cat, a species in which the very possession of color vision is not easy to demonstrate, cells showing chromatically opponent responses appear to be very rare indeed (Pearlman and Daw, 1970).

Examples of the kinds of chromatically opponent response patterns recorded from the LGN of the monkey are shown in Figure 3.22. These graphs show the average responses of a number of such cells when the eye was stimulated with spatially extended fields of monochromatic light variable in wavelength. As for the C-potentials described previously, these cells show an increase in response to some wavelengths, and a decrease to others. Because of this, the responses to some pairs of wavelengths must remain different irrespective of the relative luminances of the two. Also note that cells responding in this way show little or no response to a small band of wavelengths located roughly midway between the peaks for excitatory and inhibitory change. At that point the excitatory and inhibitory influences reflected in the cell's discharge rate are in balance. Although the examples shown in Figure 3.22 are from the monkey LGN, they are qualitatively similar to those found in other species at both ganglion cell and LGN locations. In addition to these cells showing chromatically opponent responses, other cells show either increases in firing rate to all spectral stimuli or decreases in firing rate to all spectral stimuli (see Figure 3.22). These latter units are obviously similar to the L-potentials recorded in the outer retina.

The chromatically opponent responses seen in these third- and fourth-order cells certainly reflect substantial (and probably complicated) earlier neural interactions. Nonetheless, they can be considered as approximately indexing the ratio of the effectiveness that a spectral light has on one photopigment system versus that which it has on another where the effects of the light on the two photopigment classes lead to the generation of excitatory or inhibitory changes. Consequently, the magnitude of the change in firing rate in these cells will be greatest where the ratios are largest, and the change will be zero, or roughly so, where the ratios approach one.

Figure 3.22   Average spectral response plots for six types of cells recorded from the lateral geniculate nucleus of the macaque monkey. The top four panels represent four classes of spectrally opponent cells while the bottom two graphs show spectrally nonopponent cells. The horizontal dashed line in each graph indicates the average maintained discharge rate. In each case the response was recorded for a brief flash of monochromatic light that covered a large portion of the receptive field of the cell. (Data taken from De Valois *et al.*, 1966.)

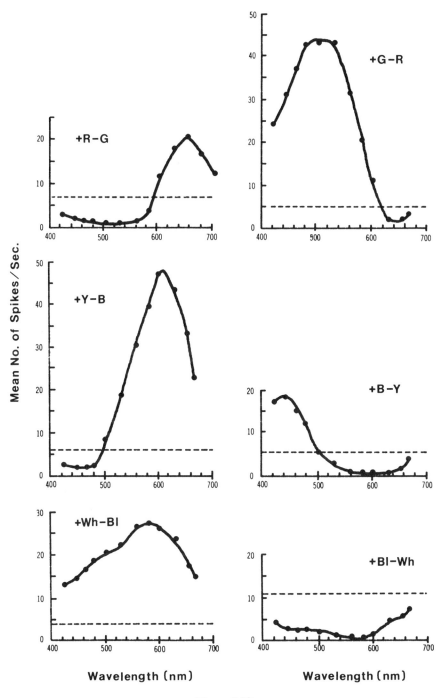

Figure 3.22

Visual systems are well designed to detect changes. This being so it is not surprising to find that these chromatically opponent cells respond briskly to stimulus change. For example, consider cells of the type labeled +R–G in Figure 3.22. These units show an increase in firing rate to wavelengths longer than about 600 nm and a decrease in firing rate to wavelengths shorter than this value. Such a response pattern could be assumed to reflect the fact that stimulation of a photopigment absorbing maximally in the longer wavelengths leads to excitation in the unit, whereas stimulation of a photopigment absorbing maximally at shorter wavelengths produces inhibition; the net response of the cell thus depends on the relative magnitudes of these two antagonistic processes. In this case, anything that increases the ratio of activation of the long-wavelength photopigment will yield increased excitation and vice versa. Thus, for example, if the light falling on the retina was abruptly changed from, say, 610 to 630 nm, the ratio of activation of the longer-wavelength photopigment would suddenly increase and so the cell would be driven to fire at higher rates. The reverse would occur if the wavelength shift were in the other direction, say from 610 to 580 nm. The point is that these cells are very sensitive to any stimulus changes which cause a change in the ratio of activation of the photopigments providing antagonistic inputs and, in particular, they are responsive to a change in the wavelength composition of the stimulus even when there is no net change in luminance. It is obvious, therefore, that these elements are particularly well adapted to transmit the information necessary to produce color vision.

The data presented in Figure 3.22 are only intended to be generically representative of the types of chromatically opponent response patterns that have been reported. A complete catalog of such response patterns thus far discovered would require much more space than the current treatment warrants. Nevertheless, it is worth inquiring in a general way if the number of discriminable chromatically opponent response patterns is limited and, if so, by what constraints. By far the most intensive analysis has been directed toward the responses recorded from ganglion and LGN cells in the macaque monkeys, therefore, that species may usefully serve as an example.

The graphs in Figure 3.22 represent one of the first attempts to categorize the various chromatically opponent responses in the macaque visual system. Because the derived categorization of response patterns of single cells in the visual system depends so heavily on the details of the stimulus conditions employed, it is worth noting that these data represent the responses of LGN cells to relatively brief, diffuse flashes of light delivered at superthreshold levels. De Valois *et al.* (1966) pointed out that in a large sample of such cells a great variety of response patterns were observed—in particular, the spectral locations of peak excitation and inhibition occupied a substantial fraction of the total spectrum. Nevertheless, it was concluded that these units could be separated into four groups, principally on the basis of the spectral location at which the responses of the cell changed from excitatory to inhibitory. Thus, the curves shown in Figure 3.22 are the averages for each of the four classes.

If one assumes that the chromatically opponent cells fall into a discrete number of classes (perhaps the four suggested in Figure 3.22), is it then possible to determine which photopigment classes are responsible for the initiation of this activity? De Valois and his associates (reviewed by De Valois, 1965) attempted to use intense chromatic adaptation to isolate the spectral inputs to such cells. Because the responses of such units represent a balance between excitatory and inhibitory influences, they are particularly susceptible to any procedure that upsets this balance. One such procedure is chromatic adaptation. In many cases when the eye is chromatically adapted, the responses of these chromatically opponent cells no longer show any evidence of spectral opponency. That is, it is possible to completely suppress that portion of the spectral response normally seen within the spectral region of the adapting light. Figure 3.23 shows the results of this procedure and gives an example to illustrate why it may be difficult to infer the nature of the underlying pigment systems without the use of a chromatic adaptation procedure.

As result of a series of chromatic adaptation experiments of this kind, it was concluded (De Valois, 1965) that the four chromatically opponent response patterns arise from pairwise antagonistic interactions of the outputs from the three classes of photopigments known to exist in macaque cones. For example, the response pattern labeled +R–G in Figure 3.22 was assumed to result from an excitatory input from a pigment having a λ_{max} at about 570 nm and an antagonistic inhibitory input from the cone pigment having a λ_{max} at about 540 nm. Similarly, +Y–B and +B–Y units were assumed to result from the antagonistic combination of signals from the cone pigment classes having peaks at about 440 and 570 nm (but see the following for a note of some dissenting opinions).

De Valois has gone on to carry out a series of experiments directed toward assessing the ways in which cells of this type might contribute to color vision. The basic paradigm involves the observation of the responses of the chromatically opponent cells when the eye is presented with the same sets of stimuli employed in various assessments of color vision—for example, comparing the responses of such a cell to two equiluminant lights of different colorimetric purities to examine how saturation might be encoded. On the basis of experiments of this sort some extensive suggestions have been made as to how these cells operate in various color vision test situations—for example, wavelength discrimination, purity discrimination, and color naming (De Valois, 1973). This approach has the virtue of providing relatively direct comparisons of physiological and psychophysical data.

An example of the kind of physiological–behavioral comparison described in the previous paragraph is illustrated in Figure 3.24, which shows a wavelength discrimination function obtained from behavioral measurements made on ground

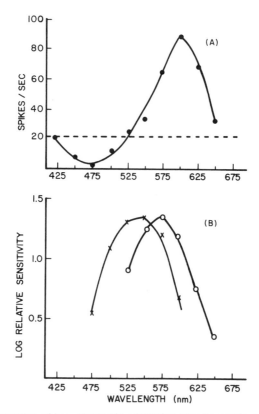

Figure 3.23 Illustration of how chromatic adaptation may be employed to determine the spectral properties of the component mechanisms underlying the responses of a chromatically opponent cell. (A) The spectral response pattern of such a cell recorded from the optic tract of a squirrel monkey when the eye was adapted to an achromatic light (dashed line: maintained discharge rate). (B) The spectral sensitivities of the two mechanisms underlying the opponent response: one was obtained in the presence of a bright red adaptation light (inhibitory component ×—×), the other in the presence of a bright blue adaptation light (excitatory component ○—○). Note that for this particular unit, the locations of peak sensitivity for both of the underlying mechanisms fall on the excitatory side in the neutrally adapted spectral response pattern and, thus, it would be very difficult to infer the spectral properties of the underlying mechanisms in this opponent response without the use of a procedure such as chromatic adaptation. (From Jacobs, unpublished data.)

squirrels (dashed line). This rodent has dichromatic color vision (see Chapter 5) and, consequently, shows most acute wavelength discrimination in the vicinity of 500 nm (near its neutral point) with relatively poorer discrimination at both longer and shorter wavelengths. In the optic nerve of the ground squirrel there are two classes of cells showing chromatically opponent response patterns. These

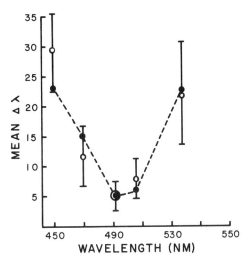

Figure 3.24 Wavelength discrimination as measured behaviorally (●- - -●) in ground squirrels plotted along with wavelength discrimination measured on a sample of units showing spectrally opponent responses recorded from the optic nerve of the squirrel. The open circles give mean discrimination values for a total of 14 units; the vertical bars show two standard deviations about the mean. See text for further discussion.

units were tested for wavelength discrimination by stimulating the eye with two rapidly alternating lights that differed in wavelength content but not in luminance. In this way it was possible to ascertain how much wavelength had to be shifted ($\Delta\lambda$) in order to produce a criterion change in the firing rate of the cell. This is closely similar to what the animal is required to do in the behavioral experiment, that is, to distinguish between two equiluminant lights of different wavelength content. The results of this test at several different wavelengths are shown by the open circles in Figure 3.24. It can be seen that there is a very good qualitative agreement between the wavelength discrimination behavior of the behaving animal and the capacities of the cells tested. It is therefore reasonable to conclude that these cells contribute the neural information needed for this color vision capacity (Jacobs, Blakeslee, & Tootell, 1981).

During the past few years the picture sketched for chromatically opponent responses in the macaque monkey visual system has become considerably more detailed. Part of the reason for this is that the more recent studies have been done in a receptive field context and thus there has been an emphasis on understanding the spatial as well as the spectral antagonisms in the responses of these cells. Not surprisingly, it turns out that spectral antagonism can arise from several different receptive field arrangements. Figure 3.25 illustrates several different receptive field classes, all of which will show chromatic opponency under the correct set of

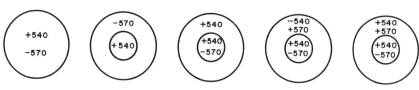

Figure 3.25 Several possible receptive field organizations each of which will show spectral opponency in response to the appropriate stimuli. In each case the large circle represents the total extent of the receptive field while the inner circle shows the boundary of the center region. The spatial distributions of inputs from two classes of cone photopigments are shown along with the sign of that input (excitatory or inhibitory).

stimulus conditions, and all of which have inputs from the same classes of cone pigments. Given these spatial complications, it is obvious that the number of discriminable classes of chromatically opponent cells will become substantially greater than the four illustrated in Figure 3.22.

A second way in which this area has been complicated is with regard to the question of the cone pigment inputs to these cells. Thus, argument has arisen as to which cone pigment classes are responsible for some of these chromatically opponent response patterns. In particular, there is uncertainty as to whether the cells labeled +Y–B and +B–Y in Figure 3.22 arise from antagonistic interactions between the 440- and 570-nm cone pigments (Abramov, 1968), between the 440- and 540-nm pigments (Wiesel & Hubel, 1966), or between several other combinations (see De Valois & Jacobs, 1981). Furthermore, there are reports suggesting that many of the cells showing chromatically opponent responses may represent inputs from all three pigment classes with two of these feeding their outputs antagonistically to those of the third pigment (De Monasterio, Gouras, & Tolhurst, 1975; Padmos & Norren, 1975).[2] Finally, both of these latter groups of investigators report that some ganglion and LGN cells that show no evidence for chromatic opponency when the eye is adapted to an achromatic light do show some chromatic opponency if the eye is concurrently exposed to a bright-colored light.

Given these recent developments can one now detail the nature of neural coding relevant to color vision seen at the level of third- and fourth-order cells in

---

[2]Authors sometimes refer to such units, those having inputs from three pigment classes, as "trichromatic." This is misleading because, as explained in the previous chapter, the term trichromatic has a precise definition in color vision. Because these cells have not been tested to ascertain whether they are in fact trichromatic, as opposed to simply having inputs from three different cone types, it is clearly inappropriate to refer to them as trichromatic. This is not mere semantic nitpicking because incautious readers can be led to quite erroneous conclusions as to how and at what levels trichromatic properties first appear in the responses of single nerve cells. For these very same reasons, one should avoid referring to those spectral locations where chromatically opponent responses change from excitation to inhibition as "neutral points."

the visual system? Perhaps not, but some general conclusions are suggested. First, it is clear that chromatically opponent response patterns are a central, and perhaps exclusive, aspect of this picture. They possess the requisite characteristics to transmit color information, and no one has seriously argued about their importance for color vision. Second, the number of discriminable response classes into which these cells can be placed depends on the nature of the stimulus conditions employed. As has been noted, if spatial configuration is added to the spectral dimension a substantial number of different response patterns emerge: De Monasterio and Gouras (1975) find 15 varieties of chromatically opponent responses among macaque ganglion cells while Spekreijse, Wagner, and Wolbarsht (1972) describe 12 types among goldfish ganglion cells. Other variations in stimulus conditions would undoubtedly lead to still further partitioning of cells according to their response patterns. This is hardly cause for alarm. Indeed, given the dramatic changes in color vision when stimulus conditions are altered, it would be surprising if cells on which this behavior is based do not also show change in their response. Insofar as their contribution to color vision is concerned, the question is not how many response classes can be discerned, but rather how these cells behave in response to stimulus conditions that are known to have important and defining implications for color vision.

E. Cortical Cells

Certainly one of the most active areas of research in mammalian neurophysiology during the past 20 years involves the analysis of the response properties of single cells in the visual cortex. Despite this research questions about color coding at this level have received only fragmentary answers. Part of the reason is because most attention was directed toward the manipulation of spatial parameters after it became clear that spatial analysis is a prime function of the cells in the visual cortex. A second reason for the relative lag in understanding the role of the cortex in color vision is that the central focus of experimental attack has been directed toward the cat, a species not noted for its excellence of color vision (see Chapter 5). Over the past several years a short list of studies on monkey visual cortex has appeared (De Valois and Jacobs, 1981). It is from these studies that some indications about cortical color signals can be gleaned.

In 1968, Hubel and Wiesel reported a study of the response characteristics of cells in the visual cortex of the macaque monkey. Among other results, they noted that some cells were "color specific" in the sense that their responses to various spectral lights, all roughly equiluminant, were substantially different. Some of these appeared similar to the kinds of chromatically opponent responses found in the monkey LGN except that the receptive fields had spatial properties characteristic of other cortical cells, that is, with plane surfaces separating regions of excitation and inhibition and consequent strong orientation preferences.

However, the most noteworthy aspect of this study was the relative paucity of cells found possessing this color specificity—only about 25% of the cells classified as "simple" according to their spatial characteristics showed any color specificity, whereas only about 10% of those cells classified as "complex" showed color properties (Hubel and Wiesel, 1968). This was particularly puzzling because, as noted previously, a very large proportion of the cells in the macaque LGN (perhaps as many as 70–80%) show chromatically opponent characteristics.

Part of the reason for the fact that the relative number of cortical cells apparently transmitting color information seemed low was that the portion of the cortex studied by Hubel and Wiesel represented the parafoveal projection area. In more recent studies directed specifically toward the foveal projection region, the proportion of cells that appear to be transmitting color information is found to be substantially higher. Thus, several studies report that as many as 50–60% of all cortical cells show some color specificity (Dow, 1974; Gouras, 1972; Yates, 1974). This representation still seems low relative to what might be expected on the basis of the LGN picture. There are at least two reasons why it might differ from expectations. First, it might simply be harder to detect a color-specific response in cortical cells than in LGN cells. Many cortical cells show little or no spontaneous activity, which obviously makes it difficult to detect inhibitory responses and so discover chromatic opponency. Second, the chromatically opponent outputs from the LGN might be utilized by cortical cells in a manner that obscures the fact that they are receiving such inputs. This possibility will be expanded on presently.

Before describing the types of potentially color-relevant signals so far detected in the monkey cortex, it is worth noting that the color selective properties of such cells may covary with other response characteristics. In particular, Gouras (1974) has reported that the more complex the spatial properties of a cortical cell, the less likely it is to show color specificity—for example, he finds that 60% of the cortical cells showing no orientational preferences have chromatically opponent properties, whereas only 17% of those categorized as complex according to their spatial properties have chromatic opponency. He suggests that this result implies the presence of separate hierarchical channels selective for either color or form (Gouras, 1972), but further evidence for this contention is lacking.

Three different response patterns have been observed among cells in the monkey visual cortex that would appear to be useful for color vision. First, everyone who has recorded from the striate cortex has found cells showing chromatically opponent responses. Some of these appear quite similar to those seen in the LGN. Others (for example, Dow, 1974; Michael, 1978) have found cells showing a double-opponent organization where, for instance, a center region of the receptive field might fire to long-wavelength stimulation and inhibit to short-wavelengths while the flanking regions show the same spectral specificity but with a reversal in the direction of the response (inhibitory, long-wavelength; excitatory, short-wavelength). Although these double-opponent cells have been seen at precortical levels in some species (as indicated schematically in Figure 3.25), it is only in the cortex that one finds them in the monkey visual system. Second, cells that only respond to a spectrally restricted range of wavelengths (for example, from 600 to 650 nm) have also been found (Yates, 1974). Since the spectral sensitivity curves for these cells appear distinctly narrower than those for photopigments, these patterns must also result from earlier opponent interactions.

In addition to these two classes of cells, a third type quite novel to the monkey visual cortex has also been described. These cells respond to flashes of light of several different wavelengths with the same type of response (De Valois, 1973). That is, of course, exactly how the L-type and nonopponent cells behave, and if no further tests were made these cells would be categorized as unimportant for color vision. However, De Valois (1973) points out that at least some of these cells will also respond to "pure hue" figures—stimuli in which there are spatial contours but no luminance differences. The nonopponent cells of the retina and LGN would be unresponsive to such stimuli so these cells, termed "multiple color cells" by De Valois, must have inputs from chromatically opponent cells in the LGN. Despite the fact that cells of this class respond to pure hue differences, they appear to show no apparent preferences for particular hue combinations; that is, they respond the same to a red line on an equiluminant green background as to the reverse arrangement. It has been suggested that these cells appear to be using color information as a means toward discriminating form (De Valois & De Valois, 1975). Gouras and Kruger (1979) report finding cells in the striate cortex of the rhesus monkey that behave in a similar fashion. It should be noted that cells of this kind behave in an analogous manner to the complex cells of the cat visual cortex, that is, the complex cells give the same kind of response to both black and white stimuli, just as the multiple color cells give the same kind of response to equiluminant lights of different wavelengths. The proportion of cortical cells in this latter class has not been specified, but their presence, along with the other types of cells described, would lead one to believe that a large majority of monkey cortical cells receive inputs from the chromatically opponent cells of the LGN—some apparently transmitting this information onward for specific color vision purposes (discrimination and identification of colors) while other cells are using comparisons of the spectral distributions of energy within the visual scene as a step toward form discrimination. Indeed, in a recent investigation it has been found that about 80% of all cells in the monkey visual cortex will respond to stimuli containing chromaticity differences but no luminance differences (Thorell, 1981). This percentage agrees well with estimates of the proportion of LGN cells concerned with color information (noted previously).

One clear direction for future analysis of the role of the cortex in color vision is to examine the properties of cells lying beyond the striate cortex but still within the visual processing chain. A few intriguing results have already been obtained. Zeki (1973, 1977) has established that the prestriate cortex of the macaque monkey contains a number of anatomically distinct areas, each receiving independent innervation and each showing a retinotopic organization. Most interestingly, Zeki has found that cells found in at least two such regions appear to be particularly responsive to the color properties of the stimulus. In one of these regions, Visual Area 4 (V4), it was found that although cells possessed a variety of spatial properties, *all* of them possessed some color specificity (Zeki, 1973). This proportion has now been revised downward somewhat (Zeki, 1978), but V4 still appears critical to the processing of color information.

One feature of the response properties of the cells found in V4 that underlines their importance for color vision is that they only respond to quite narrow spectral bands. Figure 3.26 illustrates this fact. The spectral sensitivity functions for V4 units shown have half-maximum bandwidths which average only about 20 nm. Furthermore, the spectral locations of peak sensitivities for these cells are widely

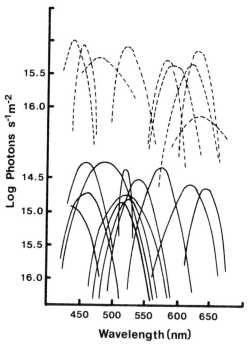

Figure 3.26 Spectral sensitivities measured on individual cells in area V4 of the monkey cortex. Units inhibited by the light falling on their receptive fields (- - -); units excited by light (—). (After Zeki, 1980. Reprinted by permission from *Nature,* Vol. 284, p. 412. Copyright © 1980 Macmillan Journals Limited.)

scattered across the spectrum, although there may be some tendency for them to be clustered in the regions of 480, 500, and 620 nm (Zeki, 1980).

Another region in the superior temporal sulcus that receives inputs from V4 also contains cells showing color specificity (Zeki, 1977). In some cases these cells also only responded to a very restricted band of wavelengths, while others showed some spectral opponency. However, despite their color specificity, most of the cells in this region do not show great spatial specificity with very little evidence for the strong orientational and directional preferences seen among cells found elsewhere in the visual cortex. In both V4 and the superior temporal sulcus, color appears to be represented topographically in the sense that cells lying in a column oriented perpendicularly to the cortical surface all tend to show the same color preferences.

F. Rod Contributions to Color Signals

The preceding sections on the neural mechanisms for color vision described how the outputs from two or more classes of cone photopigments might be processed so as to provide the requisite basis for color vision. In doing so, any potential contributions from rod signals were ignored. This is in line with the basic notion of duplicity theory, which states that color vision is associated with the operation of the cone system. However, at least two facts suggest that some attention should be paid to the rod signals as well. First, there is abundant evidence from psychophysical experiments to show that under the appropriate sets of stimulus conditions, activation of the rod system can influence color vision; second, it is known that many of the nerve fibers leading from the eye to the brain carry signals originating from both the rods and the cones.

One way in which cone and rod signals are neurally separated is by virtue of their well-known differences in sensitivity. The threshold for activation of the rod system is lower than that for activation of the cone system. Part of this appears to be due to the fact that individual rod photoreceptors can be triggered by dimmer lights than can individual cones (Fain and Dowling, 1973); part of it may be due to the presence of lateral facilitatory pathways between the rods (Fain, 1975); and part of it is undoubtedly due to the fact that rod signals are typically funneled into larger spatial pools within the retina, which would also enhance the sensitivity of the rod system to dim lights relative to that of the cones. At higher light levels, rod signals and cone signals are also separated. In 1954, Aguilar and Stiles made the fundamental discovery that as ambient illumination is raised to moderate levels, the threshold for the rod system suddenly begins to rise very rapidly, and as retinal illumination is increased still further, this threshold becomes infinite, that is, the rod system shows a saturation. At light levels above this saturation level the central visual system receives only cone-generated signals. Despite this essential ineffectiveness of the cone system at low light levels, and of the rod system at high light levels, there is still a reasonably extended luminance range over which stimuli will produce both rod and cone signals. Therefore, it is appropriate to ask how the central visual system is able to differentiate between the two types of signals.

The answer to the question of where in the visual system cone and rod signals first interact appears to be the same as that offered as the site for the earliest interactions between the various types of cones signals—at the very initial stages of retinal recording. The S-potentials in the cat retina have been shown to reflect contributions from both rods and cones (Steinberg, 1969). Interestingly, it appears that the rod input to the horizontal cells in this species occurs by virtue of passage of this signal through the cone; that is, the cones receive inputs from rods (Nelson, 1977).

The convergence of rod and cone signals has been somewhat more systematically investigated at the ganglion cell level. Gouras and Link (1966) used pairs of light flashes that activated, respectively, rods and cones in the monkey eye and varied the timing between the two flashes. They found that a signal arriving at the ganglion cell rendered that cell temporarily refractory to subsequent signals. Thus, if the cone signal arrived first it blocked the passage of subsequent rod-based signals, and vice versa. This suggests that to the extent that rods and cones respond with different characteristic latencies to a particular stimulus configuration, signals from the two types of photoreceptors are gated into the central visual system by virtue of temporal sorting at the ganglion cell level. However, Enroth-Cugell, Hertz, and Lennie (1977), on the basis of very similar experiments performed on the cat retina, have reached quite different conclusions. They found that the ganglion cell responds to the sum of the rod and cone signals that reach it irrespective of their relative timing. Whether these differences in outcome reflect species differences or are due to differences in experimental conditions is not clear. They may at any rate not hold any direct implications for color vision as the ganglion cells under study in both cases appear to have been those showing spectrally nonopponent characteristics.

Clear evidence for rod inputs into channels carrying color signals has been presented on a number of occasions. For example, Wiesel and Hubel (1966) examined the question of rod and cone contributions to responses of LGN cells in the monkey. They found that concurrent with a change from conditions of light adaptation to dark adaptation some, but not all, LGN cells showed a Purkinje shift so that the spectral sensitivity of the cell during dark adaptation was well accounted for by the spectral absorbance characteristics of the rod photopigment. Among cells showing this behavior were a number of chromatically opponent cells. Thus, for example, a cell firing to long-wavelength lights and inhibiting to short-wavelength lights when the eye was light adapted might fire to all spectral wavelengths following dark adaptation, but with maximal sensitivity at about 500 nm. This experiment and others makes it clear that cells transmitting color-relevant information under conditions of light adaptation may also transmit rod-based information during dark adaptation. What is unclear to date is how the central visual system is able to interpret the visual significance of signals from a cell transmitting both chromatically opponent and rod-based information. Presumably this might be done if the central visual system also has access to pathways that provide a monitor on the level of ambient illumination. Fortunately, there is some evidence for the presence of such pathways (for example, Barlow & Levick, 1969).

G. Short-Wavelength Cones

It has been clearly established that the short-wavelength cones (λ_{max} = 440 nm) in the human retina, often referred to as "blue cones," differ in a number of significant ways from the cones comprising the other two classes. First, the retinal distribution of these cones is unique. There are only a few such cones in the entire fovea (see Figure 3.17), and none whatsoever within the very central part of this region. Second, the blue cones frequently differ from the other two classes of cones in their susceptibility to retinal trauma and disease (Pokorny, Smith, Verriest, & Pinckers, 1979). Third, quite a number of characteristic features subserved by the short-wavelength cones have been shown to differ sharply from similarly measured capacities subserved by the other cone classes. These properties have been summarized in many places (for example, Boynton, 1979). Briefly, and most importantly, the blue cone system has significantly poorer spatial and temporal resolution than the other cone systems, and thus contributes little to the formation of visually sharp temporal and spatial boundaries. These cones also make little or no contribution to the total luminance signal.

The sources of these differences among cone systems are multiple; that is, the differences between the blue cone system and the other cone systems may arise from properties of the pigments and/or the receptors or from differences produced by virtue of the ways in which the nervous system processes the signals from blue cones relative to those from other cone systems. Whatever the sources of these differences, they are of interest from a comparative point of view because there is evidence that differences between short-wavelength cone systems and other cone systems are not unique to the human visual system. Thus far it is known that at least in the macaque monkey (Boynton & Whitten, 1972; De Monasterio, 1979) and in the cat (Zrenner & Gouras, 1979), the blue cone systems differ from the other cone systems in ways that suggest clear analogies to the differences previously detailed in the human visual system. These differences between the blue cone system and the other cone systems indicate different functional roles, and imply that these cone systems may have been evolved to satisfy different sets of needs. This issue is raised again in the final chapter.

VI. The Central Pathways for Color Vision

There is a long tradition of attempting to localize function in the nervous system; hence, it is hardly surprising that the identification of specific visual system pathways that are uniquely involved in the processing of color information has also been studied.

There are two straightforward ways to attempt to determine if a particular nervous system location is critical for color vision—either ascertain if the electrophysiological changes thought to index color vision are present at the test location or, alternatively, attempt to verify that the color vision capacity is in some way changed when the structure is rendered inoperative. The first approach involves knowing which signals are relevant for color vision, an issue discussed in detail in the previous section. One point is worth reemphasizing. The basic criterion for a usable color vision signal is that the output of the processor (neuron or neural ensemble) must be differential with wavelength changes in the face of continuously equiluminant stimuli. There has at times been confusion on this issue. In particular, one must distinguish between cone-based information that satisfies this criterion, and cone-based information that does not; that is, the mere demonstration of cone input to a location says little as to whether or not the structure is actually processing color vision information.

Experiments to determine the loci for color vision by selective inactivation of various structures have been pursued in the classical lesion–test paradigm. Using this approach, a capacity for color vision is determined both before and after surgery. Although conceptually simple, the difficulties of interpretation of these kinds of experiments are both imposing and well-known. Particular care must be paid to the possibility that the loss of a specific function is not simply due to some more global deficit—to take a somewhat overstated example, a blind subject is unlikely to perform well on a test of color vision.

From the point of view of more than three decades of lesion studies, there is no question that the geniculostriate limb of the primate visual system is immensely important for color vision. The classic experiments on this issue are those of Kluver who, in a long series of investigations, tested the residual visual capacities of monkeys that had been subjected to destruction of the visual cortex. With regard to color vision the outcome seemed to him unequivocal: "Color vision in the bilateral occipital monkey is permanently abolished" (Kluver, 1942). In addition to a loss of color vision, Kluver suggested that the light-adapted spectral sensitivity of the destriated monkey is shifted toward that generally typical of scotopic vision, the implication being that the residual vision of the destriated animal is most heavily influenced by rod-initiated activity.[3]

---

[3]Results of this kind raise the important possibility that spectral sensitivity may also be altered as a result of brain damage. If so, it can be seen that the luminance equations established for the normal

More recent experiments of the kind pioneered by Kluver frequently tend to support his conclusion. Thus, Weiskrantz (1963) found it impossible to train a rhesus monkey to discriminate between red and green papers after the animal had been subjected to bilateral removal of the striate cortex. Even though these two stimuli were not equated in luminance, and even though the animal was not tested over a very extended period of time, the fact that such a discrimination would be a trivial one for a normal animal of this species (see Chapter 5) argues that this destriated monkey had suffered a severe color vision defect. Humphrey (1970) also examined a pair of destriated rhesus monkeys. He did not test for the presence of color vision per se. However, Humphrey found that the salience of a visual stimulus was not enhanced for the subject when the stimulus was presented such that it would normally be subject to color contrast effects. This suggests again that color vision capacity in the monkey is substantially reduced when the striate cortex is removed.

Pasik and Pasik (1971) have reported a further study along these same lines that led to a somewhat different result. In this experiment, rhesus monkeys from which both the striate cortex and large fractions of Areas 18 and 19 had been removed were tested for residual visual capacities. They found that these animals were still able to discriminate a circle illuminated with red light from an otherwise identical achromatically illuminated circle. Since these investigators went to some pains to try to ensure that these two stimuli were of equal luminance, their demonstration of color vision in the destriated rhesus monkey appears quite convincing. Keating (1979) reached the same conclusion from wavelength-discrimination tests done on this same species following extensive lesions that removed all of striate cortex plus a considerable amount of tissue from the prestriate cortex.

What conclusion can be drawn from these several lesion experiments on the role of striate cortex in color vision in the macaque monkey? Given that the general capacities of the destriated monkey are far from that of the normal, it nevertheless appears that a very substantial deficit in color vision can be created by striate cortex destruction, a loss that is not explainable as a consequence of a general sensory impairment. This implies that much of the information needed for color vision ascends the geniculostriate limb of the monkey visual system. However, it also appears likely that even in the complete absence of visual cortex some color vision is possible. The experiments listed make this appear a reasonable conclusion. This further appears reasonable because it has been shown in electrophysiological experiments that neurons having an input from the fovea can be found in the monkey superior colliculus (Schiller, Stryker, Cynader, & Ber-

---

animal will not be appropriate for the brain-damaged animal. Consequently, experiments carried out using this paradigm should probably include a redetermination of the luminance equations during the testing following surgery.

man, 1974). Of course, this latter fact does not necessarily imply that the information needed for color vision reaches the midbrain, but it certainly enhances the possibility. On the other hand, given the likelihood that some information that might be utilized for color vision in the monkey reaches the midbrain, it does not appear that this information is critical for the preservation of normal color vision. For example, it is claimed that monkeys whose superior colliculi have been bilaterally ablated (80–90% complete) are able to perform color discriminations as well as these same animals were prior to surgery (Anderson & Symmes, 1969). The effects of striate cortex removal on color vision have been assessed in only a few other species besides the monkey. In one study it was concluded that albino rats were incapable of a wavelength discrimination (531 versus 612 nm) following striate cortex removal which they had been found able to perform preoperatively (Craft & Butter, 1968). Given the considerable uncertainties surrounding color vision in the rat described elsewhere (see Chapter 5), it is difficult to interpret this study.

Snyder, Killackey, and Diamond (1969) also studied the effects of striate cortex destruction, this time in the tree shrew (*Tupaia glis*). They found that these destriated animals succeeded on a variety of different color discriminations. Although no detailed analysis of color vision was performed, the discriminations which these animals were able to perform were in accord with what is to be expected from normal tree shrews (Polson, 1968). The authors argue that the difference between their results and those exemplified by the experiments of Kluver may merely reflect the fact that it is in the evolution of the higher primates that the cortex has acquired additional functions, including color vision, and the tree shrew is simply a representative of the class of visual systems in which the midbrain targets of the optic tract fibers are the more important for many functions. The correctness of their hypothesis has not yet been subjected to electrophysiological test.

Since virtually without exception all of the electrophysiological studies that have purported to discover color vision signals were directed toward the geniculostriate system, the consequences for color vision of the destruction of this system may appear to do no more than demonstrate the obvious. Nevertheless, the apparent absence of putative color signals outside of the geniculostriate system could be due simply to a lack of attention. It is certainly true, for example, that most studies done on the superior colliculus have been directed toward other concerns so that stimuli that might reveal something about color were not employed. There are a few exceptions. Thus, Marrocco and Li (1977) have reported that cells found in the superficial layers of the superior colliculus of the macaque monkey show no evidence of chromatically opponent responses. Another case in which the directionality of color information beyond the retina has been directly examined is in the ground squirrel. Michael (1968) showed that ganglion cells in the ground squirrel can be categorized into three classes—

movement sensitive, contrast sensitive, and chromatically opponent, with the latter presumed to carry color-relevant information (see Figure 3.24). In later studies, the response properties of cells in both the LGN and the superior colliculus of this animal were investigated (Michael, 1972, 1973) and a remarkable segregation of information was discovered such that only contrast and movement-sensitive cells were found in the superior colliculus, and only contrast and color-sensitive elements were found in the LGN. Unfortunately, this apparently clear picture emerging from electrophysiology is considerably muddied by a more recent lesion study (Kicliter *et al.*, 1977) which shows that thirteen-line ground squirrels are capable of wavelength discrimination following extensive destruction of the posterior neocortex, damage that should have obliterated the geniculostriate pathways. A resolution of these conflicting results is not yet available.

Because there is evidence for good color vision among the diurnal birds (see Chapter 5), it is worth noting that some attempts have been made to determine which portions of the avian central visual system are critical for color vision. Regrettably, the story is short because the evidence is all negative thus far. Neither lesions in the relay nucleus (the nucleus rotundus) interposed between the optic tectum and the telencephalon, nor lesions in the telencephalic target of the visual projection (the Wulst) appear to cause pigeons to have any great difficulty in making color discriminations (Hodos, 1969; Pritz, Mead, & Northcutt, 1970).

In addition to the experimental work on nonhuman species, there are a number of reports of pathological changes in the human nervous system that led to changes in color vision. Among the more discrete and interesting of these are cases in which cerebral damage resulted in an aphasia for color names but with no accompanying loss in color discrimination (Geschwind & Fusillo, 1966; Kinsborne & Warrington, 1964). Thus, the patient observed by Geschwind and Fusillo (1966) was quite unable to assign color names to stimuli presented to him. That this deficit did not simply involve an inability to produce color names was indicated by the fact that the patient could name from memory the colors of familiar objects (to indicate, for example, that a banana is "yellow"). At the same time, there was little or no loss of color discrimination since the patient passed with ease the standard screening tests for color-defective vision. As in most cases, the locus of the damage presumed to cause this deficit was not very well circumscribed—in this case it involved the region within the distribution area of the left posterior cerebral artery, including a destruction of the left visual cortex.

The fact that it might be possible to dissociate the information needed for color discrimination from that needed for color naming has led to the speculation that such a situation might result from a loss of the color-specific cells in V4, mentioned previously, coupled with a sparing of the so-called multiple color cells seen in Area 17 (De Valois & De Valois, 1975). The latter cells, it is argued, are able to use color as a means of discriminating form (as in a standard color vision test, the pseudoisochromatic plates) but are unable to provide the information necessary to identify particular colors. If so, the destruction of V4 (or other locations having similar properties) coupled with a sparing of Area 17 (or other regions having similar properties) might underlie the changes seen in color agnosia.

That the whole question of the cortical locus for color vision is not even as straightforward as might be suggested by the preceding is indicated by two more recent observations. The first is a case report of a patient who, as a result of cerebral infarction, suffered an extensive bilateral destruction of portions of striate and extrastriate cortex, the latter including what is believed to be area V4 (Pearlman, Birch, & Meadows, 1979). Although this patient had essentially normal vision with regard to visual acuity, eye movements, and stereopsis, he showed a very profound color vision defect. Specifically, he did very poorly on standard color vision tests, including both anomaloscope and wavelength discrimination tests. Indeed, the few wavelengths that could be discriminated were those in the immediate vicinity of 600 nm (see Figure 3.27). Neither was this patient able to name colors, although the use of color names per se was left intact. Unfortunately, again not much can be said about the exact locus of the cortical destruction in this pateint, but to the extent that Area V4 was involved it

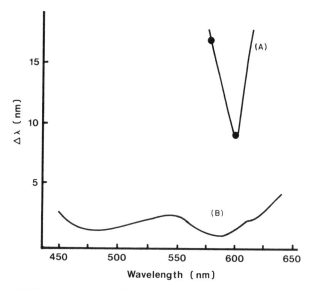

Figure 3.27 (A) Wavelength discrimination function for a patient who suffered extensive destruction of portions of the visual cortex presumed to include Area V4. (Data from Pearlman, Birch, & Meadows, 1979.) (B) The reference function represents the performance of normal human trichromats in a similar task (see Figure 2.11).

appears that serious color vision losses can be produced by the removal of this area from the human brain.

Adding still more confusion to this picture are some results involving tests of wavelength discrimination by rhesus monkeys both before and after lesions were placed in area V4 and the inferotemporal cortex (Dean, 1979). In this experiment very little if any change in hue discrimination thresholds were noted as a result of these lesions even though they should have destroyed substantial portions of those regions which Zeki's work has implicated as being important for the analysis of color information.

In conclusion, it is apparent that at the present one cannot go very far toward unequivocally specifying the central pathways for color vision. At least in primates, the geniculostriate pathway carries centrally most, but probably not all, information that may be utilized for color vision. The pathways for color vision beyond the visual cortex are presently open targets for investigation. Finally, it is worth noting that, at least in humans, it seems obvious that specification of the central pathways for color vision will have to proceed in the context of abundant evidence that the two cerebral hemispheres are differentially specialized, and thus the nature of the color vision test employed assumes prime importance. For example, there is already some evidence that in humans the right hemisphere is superior to the left when the two are separately tested in either a hue or a saturation discrimination test (Davidoff, 1976), or in a nonverbal color-matching test (Pennal, 1977).

Chapter 4

Comparative Survey of Color Vision: Nonmammalian Vertebrates

This chapter and the one that follows contain a review of studies of color vision in a number of different vertebrate species. As I have indicated earlier, although this is in no way intended to be a comprehensive survey of studies of comparative color vision, it does include information on most of those species of particular interest to color vision scientists. Lumped together in this chapter are representatives from four greatly disparate groups: amphibians, reptiles, fish, and birds.

I. Amphibians: The Frog

From his survey of the literature, Walls (1942) concluded that the amphibians have no color vision whatsoever. For at least some frog species this conclusion has now been shown to be clearly incorrect. Interest in frog color vision has been generated from several sources. First, the frog has provided a historically important preparation for studies of the nervous system and, as such, has received much attention from visual physiologists. Second, in his pioneering studies of color vision, Granit (1947) adduced some direct evidence to suggest the presence of mechanisms for color vision in the frog retina. Specifically, he found that some retinal ganglion cells showed modulator-type responses, a pattern he believed to underlie color vision (see Chapter 3). More recent neurophysiological results will be discussed presently. Finally, a relatively complete picture of the anatomy of the photoreceptors and the spectral absorbance characteristics of the photopigments in the frog retina has recently become available.

Table 4.1 summarizes the types of frog photoreceptors and their relative frequencies along with an indication of which photopigments they contain. As shown in the table, the frog retina is unusual in that it contains two classes of

Table 4.1

Photoreceptors and Photopigments in the Frog (*Rana pipiens*) Retina[a]

Photoreceptors	Relative frequency	Photopigment peak (nm)
Red rods	50%	502
Green rods	15%	432
Single cones	20%	575
Double cones	15%	Principal 575
		Accessory 502

[a] Data on receptor frequency are taken from Nilsson (1964). The photopigment measurements are based on microspectrophotometry (from Liebman & Entine, 1968).

photoreceptors designated as rods in addition to both single and double cones. The red rods are analogous to the rods typically found in other vertebrate retinas. The green rods, on the other hand, contain a photopigment having a 432-nm peak. Their classification as rod photoreceptors is based on morphology, although in terms of function they are clearly much more conelike. There are also single and double cones containing a photopigment maximally sensitive to 575 nm. The accessory member of the double cone contains the same type of photopigment as that found in the red rods. With these multiple photopigments, housed as they are in different receptors, the frog clearly has a receptor complement that could support some color vision.

Behavioral research on color vision in the frog has utilized two types of unlearned responses: a phototactic response and an optomotor response. Muntz (1962) explored the relative effectiveness of various spectral lights in releasing this phototactic response in *Rana temporaria*. His test paradigm has subsequently been used by other researchers (see Figure 2.5). To test for color preference, the frog is placed in front of two panels that can be illuminated independently. The frog jumps, either spontaneously or as a result of some small encouragement by the experimenter (in the Muntz experiment the frog was "stimulated mechanically with a small rod") toward one of the panels. The panel the frog jumps toward is recorded as that preferred between the pair presented. The basic finding is that an illuminated panel is rather consistently selected over a dark one—that is, the animal shows a photopositive response. Muntz (1962) tested the preference for various pairings of spectral lights and found that these frogs show a significantly greater preference for short-wavelength light than for other spectral lights. This basic finding has been replicated with *R. catesbiana* (Chapman, 1966) and *R. pipiens* (Jaeger & Hailman, 1971). The general nature of the outcome is schematized in Figure 4.1.

The fact that blue light is most effective in producing the photopositive re-

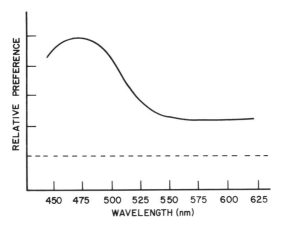

Figure 4.1 The preference for various spectral stimuli shown by frogs. The curve shows schematically the degree to which spectral stimuli are chosen when they are paired with an achromatic stimulus. This outcome is representative of that reported in several studies (see text). The distance between the solid line and the horizontal dashed line represents the degree to which spectral stimuli are preferred over a very dim achromatic stimulus.

sponse does not, of course, imply that these frogs have color vision. However, several other observations have been made that do indicate that the frog's preference is not simply based on some other feature of the stimulus, for instance, a higher sensitivity to the shorter wavelengths. These observations include, that the blue preference is clearly seen

1. whether the pairs of spectral stimuli are presented at equal physical energy (Muntz, 1962),
2. whether the pairs are equated for luminance on the basis of the frog's electroretinographic response (Chapman, 1966), or
3. when the spectral stimuli are not equated in any way (Hailman & Jaeger, 1974).

Still further convincing evidence that the expression of a blue preference requires the presence of color vision comes from experimental manipulations showing that

1. blue light is chosen by frogs over green light even when the green light is presented at several times the energy level of the blue light (Muntz, 1962),
2. 460-nm light is always preferred to 620-nm light irrespective of the relative intensities of the two (Chapman, 1966), and
3. the preference for a short-wavelength light is seen even when the frog must select against a normal brightness preference in order to express it (Hailman & Jaeger, 1974).

There is considerable evidence that not only are the middle and long wavelengths less preferred by frogs than the short wavelengths, they are, in fact, actively avoided (Muntz, 1962; Chapman, 1966; Fite, Soukup, & Carey, 1978). That is, part of the reason why short wavelengths appear to be so strongly preferred is that they have typically been tested in pairings with the longer wavelengths. Thus, if lights of different wavelengths are tested in pairwise combinations with achromatic lights that are apparently equiluminant, then frogs show positive preferences for test wavelengths shorter than about 490 nm and negative preferences for wavelengths longer than about 540 nm (Fite *et al.*, 1978). Somewhere between these two locations, presumably, frogs express no consistent preference between equiluminant chromatic and achromatic stimuli. The implication of this result is that ranid frogs have color vision that (at the least) permits them to discriminate between stimuli whose wavelength content falls roughly to the long and short side of 500 nm.

Underlying physiological mechanisms that might be used to account for the kind of color vision implied in the previous paragraph have been reported in a number of experiments. For instance, among the various types of ganglion cells recorded from the retina of *R. temporaria,* Backstrom and Reuter (1975) noted that some units show chromatically opponent responses generically similar to those described in Chapter 3. Such cells give on-responses to short-wavelength stimuli and off-responses to long-wavelength stimuli. The spectral sensitivities of these two component responses are shown by the curves in Figure 4.2. Backstrom and Reuter's analysis indicates that the on-responses arise from activation of the green rods while the off-responses reflect the activation of other cone types. Note that the response pattern for stimulus wavelengths shorter than 500 nm is qualitatively different from the responses generated by wavelengths longer than 500 nm. This difference is exactly what would be required to account for a color vision capacity sufficient to permit the preference results documented in the preceding paragraph.

Beyond the demonstration that color vision appears to be present in the frog, little more can be concluded from the work done using the phototactic response. This is due in part to a variety of complications that appear in this literature. First, Hailman and Jaeger (1974) have done a massive comparative study in which they found that not all anuran amphibians show the preference for short-wavelength light. Of the 127 species they tested, 101 were photopositive and showed the short-wavelength preference. These species were assumed to have color vision on the basis of manipulations already mentioned. Of the remaining species, a number showed a photonegative response: they chose the darker of two stimuli in a comparison of two lights. Jaeger and Hailman (1971) concluded that these species have no color vision. Apparently, even the color vision necessary for spectral preference behavior is not a universal capacity of the anurans.

Second, Muntz (1966) has shown that the nature of spectral preference be-

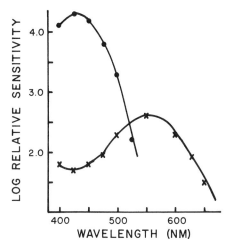

Figure 4.2 Spectral sensitivities of on (●) and off (×) components recorded from a frog ganglion cell. The curves drawn through the data points represent the sensitivities of the green rods and the cones. The small secondary rise in the short wavelengths shown by the off-component is apparently due to absorbance by the β-band of the long-wavelength cone pigment. (Taken from Backstrom & Reuter, 1975.)

havior in the frog depends both on the state of its light adaptation and on the intensity of the test stimulus. For example, whether a light-adapted frog shows a blue preference or not depends on the intensity of the stimulus. For intense stimuli a blue preference is seen, whereas for lights closer to threshold the preference behavior appears to simply mirror the spectral sensitivity of the frog. Whether this finding implies that there is a loss of color vision near photopic threshold or merely a loss of preference behavior is uncertain because clearly, the presence of a preference behavior implies the presence of color vision, but an absence of preference does not conclusively demonstrate an absence of color vision. Finally, both Muntz (1966) and Hailman and Jaeger (1974) have offered some interesting speculations about the ecological utility of spectral preference in the frog, and both have made some suggestions as to preference mechanisms. Unfortunately, there are not yet sufficient clear-cut facts about color vision in these species to make these speculations compelling.

On a variety of occasions frogs have been tested using an optomotor response (Autrum & Thomas, 1972). With this method a visual discrimination is assumed if the head movements of the animal follow the direction of movement of a striped drum surface that the frog views. The stimulus array is made up of alternating colored and achromatic stripes, the latter being variable in luminance. Obviously, in these experiments, potential luminance cues must be correctly controlled, and the stimulus stripes must be sufficiently large so that the ability of the subject to make spatial discriminations does not limit the response. Although it appears that the presence of color vision can be demonstrated by such experiments, it is not completely clear what the nature of this color vision is.

From experiments in which the optomotor response was utilized, Thomas (1955) concluded that three species of frogs (*R. esculenta, R. pipiens,* and *R. clamitans*) have trichromatic color vision. This conclusion was based on the observation that they discriminated all the spectral stimuli from achromatic stimuli. On the other hand, two other species (*R. latestei* and *R. temporaria*) appeared not to be trichromatic because they discriminated reds and blues from grays but were unable to do the same for yellows and greens (Birjkow, 1950; Thomas, 1955). These results are problematic for two reasons. First, a species that appears trichromatic by the optomotor test (*R. pipiens*) and a species that does not appear trichromatic (*R. temporaria*) in this same test appear to behave identically in color preference tests (Muntz, 1962, 1966). Second, it might be reasonable to conclude that the two species unable to discriminate yellows and greens from achromatic stimuli have dichromatic color vision except that their dichromacy is characterized by a large "neutral zone" extending all through the greens and yellows. This is not impossible, but it would be unique because in other known dichromacies the neutral points always appear to be spectrally quite discrete (see Chapter 2).

The results from both the preference and the optomotor tests argue strongly that frogs do have color vision. The results from physiological experiments and studies of photopigments suggest the same conclusion. However, details of the nature of the color vision as well as the possibility of substantial species variations remain to be established.

II. Reptiles: The Turtle

Not much is known about color vision in turtles, although a considerable number of other visual capacities have been studied (see Granda & Dvorak, 1977). Nevertheless, this group constitutes a potentially very interesting target for studies of color vision. The reason for this is because, as in the case of the frog, there is substantial information about their photopigments and photoreceptors. And, as we saw in Chapter 3, the turtle has also provided an important preparation for studies of retinal electrophysiology.

Although there are more than two hundred different turtle species, only a few of these have been studied. The retinas of turtles contain both rods and cones although the relative number of rods appears to be low. Recently, the spectral characteristics of the photopigments in the turtle retina have been measured by microspectrophotometry (Liebman & Granda, 1971). Table 4.2 shows the absorbance peaks of these photopigments for a representative freshwater turtle

Table 4.2

Photopigments in Turtle Retina[a]

	Photopigment peak (nm)	
Photoreceptors	Pseudemys[b]	Chelonia[c]
Rods	518	502
Cones	450	440
	518	502
	620	562

[a] Data from Liebman and Granda, 1971.
[b] Pseudemys is a freshwater turtle.
[c] Chelonia is a saltwater turtle.

(*Pseudemys scripta elegans*) and a sea turtle (*Chelonia mydas mydas*). Like fish, the photopigments of the freshwater turtles are based on vitamin A_2, while those of the marine species are based on vitamin A_1. Note also that, like the frog, the photopigment found in one of the classes of cones has the same spectral characteristics as that found in the rods. Thus, although the retina has four classes of photoreceptors, it contains only three different types of photopigments, at least as judged by spectral absorbance properties.

A complicating feature of the turtle retina is the presence of colored oil droplets in some cone photoreceptors. As illustrated in the previous chapter (Figure 3.6), these droplets are located in the receptor inner segments and so are in a position to alter the spectral distribution of the light reaching the photopigments in the outer segments. As I also mentioned previously, the oil droplets act as cutoff filters having high, relatively constant, absorption throughout the short wavelengths with a sharp falloff in absorption in the long wavelengths. The characteristic color of the droplet depends on the spectral location at which the falloff in absorption occurs. In turtle retinas the droplets appear to the human eye to be either red, orange, or yellow. There are also some droplets that are transparent. Those photoreceptors containing oil droplets appear to be randomly distributed throughout the turtle retina (Muntz, 1972). The magnitude of the effect of an oil droplet on the absorption characteristics of a photopigment can be very substantial. For example, Liebman (1972) has calculated that the orange oil droplet found in the retina of *Pseudemys* could produce a shift in the location of peak absorption of the 518-nm pigment out to 560 nm.

Because the effect of an oil droplet is to change the effective absorbance spectrum for the photopigment located beyond it, it is possible to imagine a color vision system based on only a single type of photopigment in conjunction with multiple classes of oil droplets. Just such a suggestion has sometimes been made for the turtle (for example, Walls, 1942). The discovery of multiple classes of photopigments (Table 4.2) appears to make this scheme unnecessary, but there is no denying that the colored oil droplets will change the spectral characteristics of the photopigments, and thus they may condition the characteristics of color vision in the turtle. Given the presence of three cone pigments along with colored oil droplets, it is likewise clear that the turtle has more than three spectral mechanisms potentially available. For example, Liebman (1972) suggests that *Pseudemys* has five possible photopic spectral mechanisms, three derived from the cone pigments located in receptors containing no oil droplets, and two more from those receptors containing the 518- and 560-nm photopigments in conjunction with colored oil droplets.

Research on turtle photopigments and retinal electrophysiology (in particular, the presence of neurons showing chromatically opponent responses) strongly suggests the potential for color vision in these animals. Unfortunately, direct studies of color vision in the turtle are very scarce. This is due, at least in part, to the fact that turtles are notoriously lethargic subjects, displaying only a very limited range of spontaneous behavior. Obviously, this makes them less than ideal subjects for psychophysical experiments.

There have been a number of investigations of visual sensitivity in the turtle. Many of these have involved the development of ingenious techniques for eliciting discriminative responses. Figure 4.3 shows several different spectral sensitivity functions derived from freshwater turtles. These were obtained from both electrophysiological and behavioral studies. It is apparent that there is substantial variation between these various estimates. Nevertheless, the fact that the turtle spectral sensitivity function is characterized by multiple sensitivity peaks is unarguable. From these several studies at least three locations of peak sensitivity can be discerned—460–480, 550, and 650 nm. How these peaks relate to the retinal filters and the photopigments, as well as to the subsequent neural processing, is not known. At the least, however, the presence of multiple peaks in the spectral sensitivity functions implies that multiple spectral channels do contribute to visually guided behavior. They are also likely to provide the substrate for some color vision, as the following studies suggest.

In an early study of color vision, Wojtusiak (1933) successfully trained turtles (*Clemmy caspica*) to discriminate colored papers from gray papers that were widely variant in relative brightness. Beyond the positive claim for the presence of color vision, he produced some evidence that these turtles discriminate between wavelengths in the region of 630 nm better than they do in other spectral regions.

There is, apparently, only one modern study of color vision in the turtle. Graf (1967) first used a flicker photometry procedure to measure spectral sensitivity in the 575–715-nm portion of the spectrum for Eastern painted turtles (*Chrysemys picta picta*). Armed with this information he next equated 625- and 685-nm lights

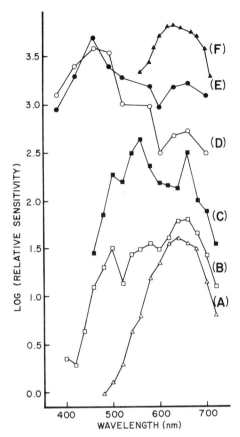

Figure 4.3 Spectral sensitivity curves derived from various species of freshwater turtles. The functions were obtained from both electrophysiological (curves A, B, C) and behavioral experiments (curves D, E, F). (After Muntz, 1972.)

so they were equiluminant for the turtle and then attempted to train five subjects to discriminate between the two wavelengths. Significant levels of discrimination were easily obtained for all animals. To show that this discrimination was not based on residual luminance cues, Graf ran a further experiment in which luminance was varied systematically (in rather large steps of 0.4 log units) around the calculated equation values. Again, these subjects showed significantly high levels of correct discrimination and thus the demonstration of color vision appears convincing. The implication from this experiment is that the turtle must have at least two spectral mechanisms operative in the 625–685-nm portion of the spectrum. Consideration of the available photopigments (Table 4.2) suggests the possibility that this discrimination might be based on the information originating from the 620-nm photopigment alone by utilizing the output from this pigment both with and without prior droplet filtering.

In summary, the behavioral work on turtles permits the conclusion that they have color vision. Beyond that, little can be added at the present time. Nevertheless, the evidence on photopigments, retinal filters, and neural coding can be taken to suggest that turtles not only have color vision but that it may very well be a significant feature of their visual capacities.

III. Fish: The Goldfish

Interest in fish color vision is long standing. By 1942, when Walls published his survey of vertebrate vision, there were already a substantial number of investigations dealing with color vision in fish (see Walls, 1942, pp. 472–490). In general, many species of fish appear to have good color vision (see Autrum & Thomas, 1972). Various of the cyprinid species (like goldfish, carp, and tench) have received particular attention from the electrophysiologists. The goldfish (*Carrasius auratus*) provides a representative subject for our consideration.

The retinas of the goldfish and other cyprinids contain both rods and cones. The rod photopigment is a porphyropsin having peak sensitivity at 522 nm. The cone pigments have been studied in recent years with the microspectrophotometer (Liebman & Entine, 1964; Marks, 1965; Harosi, 1976). Based on a sample of about 100 photoreceptors, the difference spectra for the cone pigments were found to fall into three discrete classes showing absorbance peaks at 455 ± 15 nm, 530 ± 5 nm and 625 ± 5 nm (Marks, 1965). The goldfish retina, like that of the frog, has both single and double cones. Of the latter, almost all of the pairs appear to contain the 530- and 625-nm photopigments (Marks, 1965). Given this complement of photoreceptors and photopigments, along with abundant evidence for the presence of spectrally opponent units within the retina (Spekreijse *et al.*, 1972), it is not surprising that most writers have simply assumed that the goldfish must have trichromatic color vision. This assumption is almost certainly correct.

Most of the behavioral work on the goldfish has utilized some kind of a forced-choice discrimination task. One exception to this involved the conditioning of an autonomic response—heart rate (McCleary & Bernstein, 1959). In this experiment goldfish were trained to discriminate between red and green stimuli. These stimuli were equated to be equally bright to the human eye. Obviously, given the difference in cone photopigments between man and goldfish this was a less than optimal procedure. At any rate, after acquisition of this discrimination the fish were tested for generalization of the learned response to both "bright" and "dim" red and green stimuli. The result was that the fish generalized according to hue; that is, if the subject had been trained to respond to the red stimulus it showed the same response to both the bright and the dim red stimuli in the generalization test, but did not show the response to the green stimuli. This result would be hard to explain if the goldfish lacked color vision. A few years later, Yarczower and Bitterman (1965) examined wavelength discrimination and wavelength generalization in the goldfish. Because no attempts were made to take into account possible luminance cues, the results of this experiment provide no compelling conclusions about goldfish color vision.

Muntz and Cronly-Dillon (1966) used a two-choice discrimination apparatus to test for color discrimination in the goldfish. The stimuli were painted panels (blue, green, and red), each of which was presented at several different illuminance levels. Six groups of subjects were tested on all possible pairings of these stimuli with the relative brightnesses of the pairs varied over a wide range so as to try and make brightness an irrelevant cue. All of their subjects learned the discrimination rapidly and, consequently, the experimenters concluded that goldfish must have good color vision. They also concluded that goldfish color vision is trichromatic. However, as Yager and Jameson (1968) subsequently pointed out, because some human dichromats could have also performed credibly on these spectral discriminations, the conclusion that goldfish are trichromats is not strictly justified by this experiment.

Yager (1967, 1974) has carried out detailed studies of spectral sensitivity and colorimetric purity discrimination in goldfish. He tested fish in a two-choice discrimination task where the stimuli intended for discrimination were projected onto response levers that the fish were trained to strike in order to secure food reinforcement. A novel aspect of this experiment was that the fish also had to strike a third lever in order to initiate a test trial. The advantage of using an "observing response" of this kind is that it places the subject in a more or less standard location relative to the stimuli on each trial, and it also assures that the subject is prepared to attend to the test stimuli. After first obtaining spectral sensitivity functions and thus being able to control for possible luminance cues, Yager measured the ability of the fish to discriminate colorimetric purity at a number of different test wavelengths. A test trial consisted of adding monochromatic light to one of the two stimuli. Achromatic light was simultaneously added to the other stimulus in order to keep the luminances of the two equivalent. The fish was required to select the stimulus containing the monochromatic light; and thus the energy of the monochromatic light required for threshold detection was determined after a number of trials. The results of this procedure for a wide range of test wavelengths are shown in Figure 4.4.

The data shown in Figure 4.4 are presented in the same way as is conventionally done for human saturation discrimination data—that is, the reciprocal of threshold colorimetric purity is plotted for each test wavelength. Judged in this context, the experiment provides some strong inferences about goldfish color vision. First, the fact that all spectral stimuli were discriminated from equally

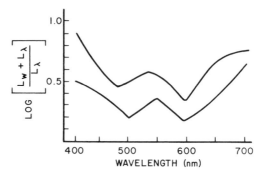

Figure 4.4 Spectral saturation discrimination by goldfish. The lines enclose the range of results for three subjects. (After Yager, 1967.)

luminant achromatic lights implies that goldfish color vision must be trichromatic. This conclusion is reinforced by results from a more recent experiment that employed a similar type of measurement (Shefner & Levine, 1976). Second, the form of the function obtained by Yager also suggests some interesting conclusions. Although not totally unambiguous, it appears that the spectrum contains three regions of relatively high saturation for the goldfish—at the spectral extremes and in a zone extending from about 510 to 535 nm. Locations of low saturation occur at about 490 nm and much more clearly at about 600 nm. The form of the function is analogous to that measured for normal human trichromats, having two locations of low saturation. It differs dramatically, however, (1) in the degree to which the spectral extremes appear saturated (the short wavelengths appear much more saturated to the human eye), and (2) in the location of one of the minima. For the goldfish this minimum is at about 600 nm whereas for the human it is at 570 nm. Much of this latter difference can undoubtedly be traced to the differences in the absorbance spectra of the cone pigments in the two species (Yager, 1967); some of it may also be due to differences in the manner in which the two species have been tested.

   In an earlier discussion of the neural mechanisms for color vision, it was noted that rod signals might be contrasted with cone signals under some sets of stimulus conditions to produce a neural code for color. In a recent study of wavelength discrimination, Powers and Easter (1978) argue that just such a conjunction must occur in the visual system of the goldfish. Specifically, they found that goldfish were able to successfully discriminate between 636- and 532-nm lights at intensities that were only one log unit above absolute visual threshold. They suggested that such a discrimination might be based on qualitative differences in the signals originating from the rods and the long-wavelength sensitive cones. Physiological evidence to support this conclusion is not yet available.

IV. Birds: The Pigeon

Although apparently no one has ever seriously questioned the presence of color vision in the diurnal birds, its details and mechanisms remain largely unresolved. Because of its behavioral tractability and its suitability as a physiological preparation, the pigeon (*Columba livia*) has been the subject of a substantial number of investigations. Whether or not it can be taken as a representative example of avian color vision remains to be established.

The anatomy of the pigeon retina shows the following picture: both rods and cones are present, although a large preponderance of the photoreceptors are cones. Among the latter, both single and double varieties are found. The retina is conventionally divided into two regions on the basis of its macroscopic appearance. One of these, the so-called "red field," is elliptical in shape. This region occupies roughly the posterior dorsal quadrant of the retina and encompasses much of the bird's binocular field of view. The yellow field makes up most of the rest of the retina. As in the turtle, many of the cones in the pigeon retina contain highly colored oil droplets. It is the differential distribution of the different types of droplets that accounts for the characteristic appearance of the red and yellow fields. The absorbance characteristics of the oil droplets found in the pigeon retina have been measured on several occasions. One such series of measurements is summarized in Figure 4.5. As can be seen, there are several types of oil droplets. Those found in the red field (Figure 4.5B) are red, orange, or yellow in appearance. These droplets are most frequently located in single cones, as double cones and rods are relatively rare in this region. The yellow field (Figure 4.5A) contains a higher proportion of rods and double cones. It too has red, orange, and yellow droplets. Although both fields contain droplets carrying the same color designations, it is apparent that the absorbance curves are generally displaced toward the short wavelengths for those droplets found in the yellow field. The maximum light transmission to short-wavelength irradiation amounts to 10% or less for all classes of oil droplets. Another difference between these two retinal regions is the presence of very small oil droplets (less than 0.2 μm in diameter) in the inner segments of cones in the red field. These small droplets are red in appearance, the same as that of the large droplet found in the photoreceptor (Pedler & Boyle, 1969). In line with arguments advanced previously, this rich variety of oil droplets obviously suggests the possibility of a color vision mechanism in the pigeon retina.

Agreement on the identification of the photopigments in the pigeon retina remains elusive. Table 4.3 summarizes a number of estimates of the absorption characteristics of pigeon photopigments. The rods contain a typical vitamin A_1-based photopigment having maximum absorbance at 503 nm. Claims have been entered (Table 4.3) for cone photopigments whose peak sensitivities fall

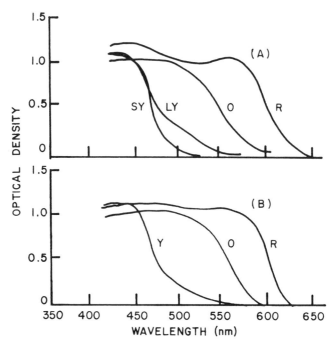

Figure 4.5 Absorbance measurements of oil droplets in (A) the yellow field and (B) the red field found in the pigeon retina. The curves are means for measurements made on several oil droplets. Abbreviations: red, R; orange, O; large yellow, LY; small yellow, SY. (Data from King-Smith, 1969.)

into at least six different spectral regions. In the face of all these claims it is obviously risky for anyone to draw any very strong conclusions as to which are correct. However, a few comments may be offered. First, there is substantial agreement that one of the pigeon cone pigments has peak absorbance at 560 to 570 nm. Not only has this pigment been repeatedly found in direct measurements, it also corresponds closely to the location of peak photopic sensitivity in the pigeon (Blough, 1957; Romeskie & Yager, 1976a). Beyond that, one would be most inclined to place greatest weight on those measurements that were made the most directly, in this case by MSP. However, even here caution is required because of the technical difficulties inherent in these measurements, some of which were alluded to earlier, and because of the fact that MSP is based on a sampling procedure so that some classes of photopigments might be missed entirely if they are present in small numbers of photoreceptors, or if they are spatially segregated within the retinal mosaic. Given all of these cautions, the recent measurements made by Bowmaker (1977) should probably be accorded the greatest credence.

Table 4.3

Estimates of the Photopigments Found in the Pigeon Retina[a]

Receptor type	Peak (nm)				Source
Rods				500	Liebman (1972) (MSP)
				502	Bridges (1962) (extracts)
				503	Bowmaker (1977) (MSP)
				507	Govardovskii & Zueva, (1977) (electrophysiological)
Cones	400				Graf & Norren (1974) (electrophysiological)
	415		480		Norren (1975) (electrophysiological)
	413	467		514	Govardovskii & Zueva, (1977) (electrophysiological)
				544	Bridges (1962) (extracts)
				560–575	Liebman (1972) (MSP)
	461			567	Bowmaker (1977) (MSP)

[a] The tabled values represent measurements based on either photopigment extracts, MSP, or various electrophysiological indices of gross retinal activity. The results are arranged in columns to indicate that several different classes of photopigments are present in this retina.

Under the assumption that there is only one type of cone pigment in the pigeon retina, a number of investigators have sought to show how this single pigment in conjunction with various oil droplets might provide the basis for a wavelength-discriminating mechanism. Given the wide variety of droplet absorbance functions (Figure 4.5), it is apparent that the effective absorbance function for a single photopigment might assume a variety of different forms depending on which of the droplets it is paired with. However, because the droplets all act as effective high-pass filters, all they can do in conjunction with a single photopigment is to move the location of peak absorbance greater or lesser distances toward the long wavelengths. Thus, if peak sensitivity of the cone pigment is at the location suggested by the measurements of Liebman (1972), at 560 to 575 nm, then in conjunction with a red droplet it would show peak absorbance at 625 nm; in conjunction with an orange oil droplet absorbance would be maximal at 585 nm, and so on. Obviously, with a single pigment peaking in the region of 560 nm, even in conjunction with a number of different oil droplets, color vision would be extraordinarily poor (or probably absent entirely) in the short wavelengths.

The possibility that the pigeon retina contains only a single cone pigment has, in any case, disappeared. Evidence that there may be a cone photopigment with considerable sensitivity below 500 nm comes from several sources. First, pigeon spectral sensitivity functions show a clear secondary peak in the short wavelengths, whether they are determined behaviorally (Blough 1957; Romeskie & Yager, 1976a) or electroretinographically (Graf & Norren, 1974). This secondary peak would be hard to account for except on the basis of another photopigment absorbing maximally in the short wavelengths. Furthermore, it is apparent that ultraviolet light can be used to directly influence the visual behavior of the pigeon at photopic light levels (Wright, 1972a; Emmerton & Delius, 1980; Kreithen & Eisner, 1978). Again, this would probably not be expected to occur if the retina contained only a 560-nm pigment.[1] Finally, Graf and Norren (1974) and Norren (1975) have carried out studies of the effects of chromatic adaptation on the spectral sensitivity of the pigeon electroretinogram, and they present evidence for two mechanisms in the short-wavelength end of the spectrum, one peaking at 400 to 415 nm and one at about 480 nm. The former, in particular, seems fairly clearly to reflect a short-wavelength photopigment system. In sum, in addition to the cone pigment measured directly by extraction procedures,

---

[1]The possibility that these short-wavelength effects might result from absorbance by the beta band of the long-wavelength photopigments cannot be completely discounted (see Chapter 3). Additionally, Goldsmith (1980) has recently found that three species of hummingbirds are also capable of discriminating light from the near-ultraviolet portion of the spectrum. He suggests that this high short-wavelength sensitivity implies a richness in avian color vision that places it beyond that found among the primates.

showing peak sensitivity at about 560 nm, there is growing evidence from both direct and indirect measurements for other pigment systems with maximal response in the shorter wavelengths. Very likely there are at least two of the latter.

There is very little information about the neural mechanisms that might underlie color vision in the pigeon. Yazulla and Granda (1973) reported that some units recorded from the thalamus (the nucleus rotundus) of the pigeon gave spectrally opponent responses. Only one essential pattern was seen, and it was roughly similar to that shown by the Y/B cells illustrated in Figure 3.18. However, the total number of such cells was quite small as they were found to make up only about 15% of all cells.

The first behavioral experiments on pigeon color vision were completed nearly 50 years ago (Hamilton & Coleman, 1933). These investigators trained birds in a Lashley-type jumping stand to discriminate between various pairs of chromatic stimuli. Under the assumption that the pigeon spectral sensitivity function was similar to that of man, they varied the relative luminances of the spectral stimuli sufficiently to "eliminate any possibility of reactions to position or brightness and to convince ourselves that without doubt the animal was reacting to wavelength alone." This accomplished, they measured wavelength discrimination over a spectral range from 460 to 700 nm and found that pigeons could discriminate wavelengths throughout this range. Their conclusion was that these birds are trichromatic.

Wavelength discrimination in the pigeon has also been measured in a series of more recent studies. The outcomes of two of these experiments (Blough, 1972; Wright 1972b), along with the results of Hamilton and Coleman (1933), are summarized in Figure 4.6. All of these studies agree that the pigeon is able to discriminate wavelengths over a spectral range broad enough to be clearly suggestive of trichromacy. Beyond that, as Figure 4.6 shows, there is considerable uncertainty about the precise form of the wavelength-discrimination function. There is good agreement that pigeon wavelength discrimination is most acute at about 600 nm—the studies summarized in Figure 4.6, as well as an experiment by Bloch and Martinoya (1971), all found a minimum in the wavelength-discrimination function at 590 to 600 nm. However, for the shorter wavelengths the form of the function does not appear to be so clearly defined. Three studies also found a minimum at about 500 nm (Hamilton & Coleman, 1933; Bloch & Martinoya, 1971; Wright, 1972b). Whether or not there is a third minimum, and where its location might be, is uncertain. From the viewpoint of technique and thoroughness, the data of Wright (1972b) should probably be accorded the greatest weight—although the function he measured is essentially flat at wavelengths shorter than about 580 nm, there are suggestions of three locations of best discrimination, at 600, 540, and 500 nm. Emmerton & Delius (1980) have reported additional measurements suggesting that there may be a fourth minimum in the pigeon wavelength discrimination function, in the near ul-

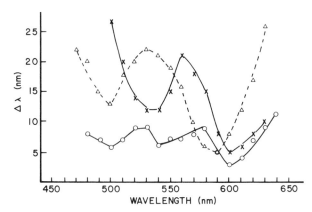

Figure 4.6 Wavelength discrimination functions for the pigeon. The data from Blough (1972) were derived indirectly from a study of wavelength-generalization gradients. The results from the other two studies represent direct measurements of wavelength discrimination functions. △- - -△, Hamilton and Coleman (1933); ○—○, Wright (1972b); ×—×, Blough (1972).

traviolet at 365 to 385 nm. Taken together, all of these studies strongly support the claim of excellent color vision for the pigeon even though the exact form of the wavelength-discrimination function cannot yet be said to have been conclusively established. An obvious next step would be to evaluate the possibility that the discrepancies between these various measurements might result from the fact that different regions of the retina have been tested in the various experiments.

The idea that differences in the outcomes of color vision tests on pigeons might be due to differences in the regions of the retina being tested in the various studies is made plausible by recent behavioral measurements of spectral sensitivity. Martin and Muntz (1979) conducted an experiment in which a headgear mount was designed to be worn by pigeons while they were performing discrimination tasks. Within this headgear were two lights, positioned so that one could be used to illuminate a small region within the yellow field of the retina, while the other could illuminate a like-sized region within the red field. The resulting spectral sensitivity functions obtained from these two portions of the retina are shown in Figure 4.7. Although there is little difference in these sensitivity curves for wavelengths longer than about 550 nm, at shorter wavelengths these birds were substantially less sensitive for lights that fell within the red field. These clear differences in spectral sensitivity suggest that the two retinal locations yield differences in the details of the color vision.

Because the pigeon is a popular subject for experiments on learning in which different spectral stimuli are often used as discriminative stimuli, Wright (1978) has utilized the wavelength discrimination results to provide a means of easily

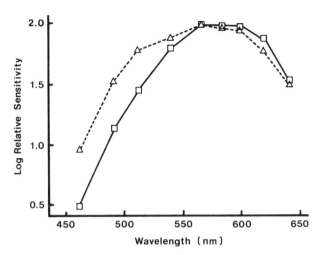

Figure 4.7 Mean spectral sensitivity functions determined in behavioral experiments in which the stimuli were imaged onto either the red (□—□) or yellow (△- - -△) fields of the retina. (Data taken from Martin & Muntz, 1979.)

determining the size of equally discriminable wavelength steps throughout the spectrum. This has the practical utility of permitting an experimenter to avoid the confounding of variations in wavelength discrimination capacity with other performance measures.

In addition to studies of wavelength discrimination, there is also some evidence from behavioral experiments bearing on spectral saturation discrimination for the pigeon. Schneider (1972) conducted experiments in which pigeons were trained to assess the degree of similarity between various lights. Among other things, it was found that pigeons were more likely to categorize spectral lights between 542 and 568 nm as the "same" as a white light than they were for lights taken from other spectral regions. The implication is that this spectral region appears most desaturated to the pigeon.

A somewhat similar experiment was done by Blough (1975). She tested pigeons to find out how well they generalized between a white light and various monochromatic lights, asking, in essence, which spectral stimuli appeared "most like" an equiluminant white light. Judged in this way, the spectral region of minimal saturation for the pigeon was from 575 to 625 nm, and there was a pronounced increase in saturation below about 550 nm.

Finally, Romeskie and Yager (1976b) used a two-choice discrimination task to obtain measurements of the photochromatic interval for pigeons. This measure represents, for any given wavelength, the threshold difference between the discrimination of the presence of a light and the discrimination of its chromatic character. In the human the size of the photochromatic interval correlates with the degree to which a spectral light appears saturated. According to this analysis, spectral saturation for the pigeon was minimal at 600 nm, and there was also some hint of secondary regions of lowered saturation at about 657 and 450 nm (Romeskie & Yager, 1976b). In these three studies rather disparate techniques were used, and they do not agree very precisely in outcome. Nevertheless, it does appear that for the pigeon the spectrum is minimally saturated around 600 nm. This spectral location is also where wavelength discrimination is most acute (see Figure 4.6). This sort of correspondence between wavelength discrimination and spectral saturation is much like that seen in human color vision.

In man, the hue of a spectral light depends both on its wavelength and on its intensity. Specifically, if the wavelength of a light is held constant, then the hue of that light changes as the intensity of the light is increased or decreased. These changes are called the Bezold-Brücke hue shift; their characteristics and theoretical significance have been much studied in human subjects. Wright (1976) has attempted to determine if these shifts are also characteristic of pigeon color vision. The context for assessing this possibility was a wavelength discrimination task in which two equiluminant lights were projected onto two halves of a bipartite field and the pigeon was trained to give a response indicating whether the two stimuli appeared the same or different. After the birds had received very extended training on this task, on an occasional test trial, the luminance of one of the stimuli was made higher or lower than that of the light with which it was paired. Comparisons of the birds' performance on these occasional trials with its performance on trials where only equiluminant lights were presented permitted the experimenter to assess whether, and how much of, a change in wavelength would have been required in order to account for the bird's performance on trials where the two stimuli were varied in luminance. Wright found that for the pigeon, hue does change as a function of stimulus luminance alone, that is, that a Bezold-Brücke shift is also characteristic of the color vision of this species. Beyond demonstrating the presence of a luminance-dependent hue shift, Wright (1976) noted that at four wavelengths (530, 550, 600, and 630 nm) hue does not change as luminance is raised or lowered. Spectral regions such as these are called invariant points. Three such points, at about 480, 505, and 575 nm, are typically found in normal human trichromats. Although the significance of these locations continues to be much debated by color vision theorists, for the human, at least two of these locations correspond to the regions of most acute wavelength discrimination. This type of correspondence would also seem apparent for the pigeon—the location of one of these invariant points (600 nm) is also the region in which this species shows most acute wavelength discrimination (Figure 4.6).

The form of the human wavelength-discrimination function has often been directly related to the spectral boundaries between hues; that is, wavelength discrimination is best just at those locations where hue changes most rapidly, as at the transitions from yellow to green or from yellow to red. On the grounds that it might be possible to gain some information about spectral appearance from nonhuman subjects, Wright and Cummings (1971) conducted an ingenious experiment on "color naming" in the pigeon. The paradigm employed a three-key, matching-to-sample task in which the pigeon was presented with three spectral stimuli and was trained to peck at the wavelength appearing on one of the side keys that matched the wavelength presented on the center key. For example, if a 570-nm light appeared on the center key with 510- and 570-nm lights on the two side keys, then the bird was reinforced for pecking on the side key that was illuminated with the 570-nm light. Once the birds were thoroughly trained in this task for several different wavelengths, on some trials a test (probe) stimulus appeared on the center key. Thus, if the bird had been trained with both 572- and 512-nm lights appearing on both center and side keys, on test trials the 572- and 512-nm lights were presented on side keys and a probe wavelength (say, 540 nm) was presented on the center key. The experimental question was, given the 540-nm stimulus, does the bird match it to the 572-nm or the 512-nm light? Because these wavelength differences are much greater than those required for mere discrimination (Figure 4.6), it was argued that the bird's response must be based on placing two of the stimuli in the same category. Repetition of this procedure for a large range of test wavelengths resulted in the categorization functions shown in Figure 4.8.

The implication of the data shown in Figure 4.8 is that the pigeon is able to consistently group adjacent wavelengths into categories (possibly hues). Furthermore, the transitional boundaries between these categories are clear-cut. As shown in Figure 4.8, Wright and Cumming replicated the basic experiment with a second set of training wavelengths. The results from these two experiments are similar enough to imply that the hue categories formed by the pigeons are not

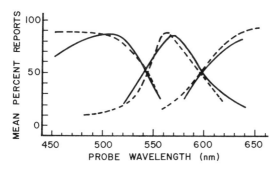

Figure 4.8 Categorization of spectral wavelength by the pigeon. Dashed and solid curves show the results from two different experiments. See text for further details. (After Wright & Cummings, 1971. Copyright © 1971 by the Society for the Experimental Analysis of Behavior, Inc.)

simply artifacts of the particular wavelengths used. There are two other interesting aspects of this experiment. First, the transitions between categories for the pigeon are at about 540 and 595 nm. These wavelengths correspond to two of the three minima in the wavelength-discrimination function (Wright, 1972) suggesting that the explanation for the wavelength-discrimination function on the basis of spectral appearance holds for the pigeon as well as it does for man. Second, these categories are quite different from those obtained in color-naming experiments with humans. Thus, as Wright and Cumming point out, wavelengths on either side of 540 nm fall into different categories for the pigeon, whereas they clearly do not for the normal human observer tested on a similar task. Among other things, this result strongly emphasizes how misleading it may be to use human hue designations to describe color vision in nonhuman species.

It is obvious that there have been a large number of exceptionally high-quality behavioral studies done on the pigeon. What can we conclude about their color vision? First, it is clear that this bird possesses excellent color vision. All of the quantitative measures obtained indicate that the capacity is an acute one—for example, stimuli differing by no more than a few nanometers are clearly discriminable to this species. Second, in terms of formal classification it seems clear that the pigeon must have at least trichomatic color vision. No evidence of a dichromatic neutral point has ever been reported, and in many other ways the indices of color vision in the pigeon roughly resemble those shown by human trichomats. However, at the same time, it is not possible to be certain that only three spectral dimensions are required to account for color vision in the pigeon. That is, the pigeon could be at least tetrachromatic. Indeed, the presence of three classes of cone pigments and several types of oil droplets leads Bowmaker (1977) to suggest that both the red and yellow fields of the pigeon retina contain six classes of cones showing different spectral absorbance characteristics. On these grounds alone the possibility of tetrachromatic or perhaps pentachromatic color vision exists. What is clearly needed is a thorough investigation of the limits of color mixing by the pigeon. The one study of this kind that has been done so far suggests that the pigeon does not produce dichromatic color matches over a range of the spectrum where the normal human trichromat does produce such matches (Jitsumori, 1976). One possible explanation for this result is, of course, that the pigeon does not have trichromatic color vision.

V. Colored Oil Droplets and Color Vision

The utility of retinal oil droplets remains an enigma. As indicated in Chapter 3, one persistent suggestion has been that the oil droplets serve to produce multiple spectral channels and that they therefore constitute a prime mechanism for color vision. This idea gained further support when early attempts to measure the cone pigments in bird retinas seemed to indicate that only one pigment was present. However, as we have just seen, there is now evidence that the retinas of diurnal birds contain more than one photopigment—very likely three. If this is true the colored oil droplets in the photoreceptors are not strictly required for color vision. What then might be their role in color vision?

From an experimental standpoint the role of oil droplets could be evaluated by comparing the vision of some species both with and without its retinal oil droplets. In at least one species of bird this somewhat unlikely prospect is now realizable. Meyer, Stuckey, and Hudson (1971) showed that Japanese quail (*Cortunix cortunix japonica*) raised on a carotenoid-free diet have only colorless oil droplets. Clearly, then, in order to evaluate the role of colored oil droplets in color vision what needs to be done is to compare the capacities of normal birds and those in which the droplets have been rendered colorless. This has not yet been accomplished in a completely satisfactory way, although there are some studies of color vision in the Japanese quail both with and without colored oil droplets (Kovach, Wilson & O'Connor, 1976; Duecker & Schultze, 1977; Wallman, 1979). Unfortunately, although it is clear that this species has color vision, its characteristics have not been delineated in the same extensive way that they have for the pigeon. Furthermore, the color vision in the birds of this species who lack colored oil droplets has also not been evaluated very extensively. Nevertheless, it appears clear that those birds without colored oil droplets still have some color vision, although it may be poorer than that of normal birds. For instance, in one study utilizing an optomotor response, Wallman (1979) found that although carotenoid-deprived Japanese quail still discriminated between spectrally broadband red and green lights, their vision appeared to be more influenced by luminance differences in the stimulus pattern than were normal birds similarly tested.

Although it would be unwarranted to conclude that the colored oil droplets found in the retinas of a large number of species have no role in color vision, it now seems highly unlikely that they are a primary mechanism. The possibility that they serve as accessory mechanisms to sharpen color discrimination still has some merit. At any rate, the experimental tools to resolve the issue once and for all would now appear to be at hand.

Chapter 5

Comparative Survey of Color Vision: Mammals

Walls' (1942) opinion that color vision is "by no means widespread among mammals" has been propagated so widely that his view still seems to dominate most discussions of mammalian color vision. Walls' conclusion was based on a consideration of the experimental data available in 1940, and on the reasonable contention that because strong diurnality is relatively uncommon among mammals, it is therefore unlikely that many of this group would have evolved good color vision. Studies done over the past few years have greatly expanded the view of mammalian color vision, and it can now be concluded that color vision, at least in its strict definitional sense, is a good deal more widespread among mammals than was previously supposed. Nevertheless, it is likewise clear that there are wide variations in the presence, quality, and importance of color vision among various mammalian groups. As with the nonmammalian vertebrates, I shall examine the literature on color vision for representative mammalian species.

I. Rodents

Because rodent species span the entire range from strongly diurnal to strongly nocturnal in behavior, they comprise an interesting group from a color vision viewpoint. Perhaps a reasonable perspective can be achieved by a consideration of the squirrels, species that are generally strongly diurnal, and from a consideration of the rat, a strongly nocturnal rodent of interest primarily by virtue of its widespread use as a common laboratory subject.

A. Squirrels

Consider first the ground-dwelling sciurids (ground squirrels, prairie dogs, and the like). These animals are clearly diurnal in habit, feeding and moving about only during the daylight hours. Based on anatomical, physiological, and behavioral observations, these animals were for years believed to have all-cone retinas. Now, however, it has become clear that most, perhaps all, of the ground-dwelling sciurids do have mixed retinas containing both rods and cones (Green & Dowling, 1975; Jacobs, Fisher, Anderson, & Silverman, 1976; West & Dowling, 1975). The number of rods in these retinas is small, probably not making up more than 5–10% of the total number of photoreceptors. This mixture of rods and cones makes the retinas of these animals unique among the mammals.

With regard to the photopigments present in the retinas of ground-dwelling sciurids, it is obvious that the story is not yet complete. The minority population of rods appear to contain a typical mammalian rhodopsin having peak sensitivity at about 500 nm (Green & Dowling, 1975; Jacobs et al., 1976).[1] In 1964, Dowling reported densitometry measurements carried out on the retina of the thirteen-lined ground squirrel (*Spermophilus tridecemlineatus*). He found evidence for the presence of a single type of photopigment having a sensitivity peak at 523 nm. Although there are no other direct measurements of photopigments in these animals, there have been a number of studies done on the electroretinogram of the ground squirrels (reviewed in Jacobs & Yolton, 1971). Many of these studies report finding spectral sensitivity functions characterized by multiple sensitivity peaks. This latter result clearly implies the presence of more than one photopigment in the ground squirrel retina. Considerably more persuasive in this regard are studies of the responses of single cells in the visual systems of thirteen-lined and Mexican ground squirrels (*S. mexicanus*), which indicate that there are likely two cone photopigments in the retinas of these animals having absorbance peaks at about 460 and 525 nm (Michael, 1968; Tong, 1977; Gur & Purple, 1978). Single unit studies done on the California ground squirrel (*S. beecheyi*) also indicate the presence of two cone pigments, in this case with peaks at 525 and 440 nm (Jacobs and Tootell, 1981). In conjunction with recent results on rods, it appears that the ground-dwelling sciurids have both a feeble scotopic system and a photopic system based on the operation of at least two classes of photopigments. The latter clearly suggests the potential for some color vision.

Beyond providing information on possible cone pigments in the ground squirrel retina, investigations of the electrophysiology of the ground squirrel visual system have also yielded some results about possible neural mechanisms associated with color vision. Unfortunately, the conclusions reached in the various studies are somewhat inconsistent. On the one hand, several investigators have

---

[1]There is some evidence that there may be considerable individual variability among ground squirrels in the possession of a viable scotopic visual system. Specifically, in a survey involving the recording of gross electrical potentials from the retinas of California ground squirrels, it was discovered that about 30% of the animals tested yielded no scotopic signals even though they were examined under conditions that should have been optimal for eliciting such activity (Jacobs, Tootell, Fisher, & Anderson, 1980).

reported finding single units showing chromatically opponent responses with spectral properties indicating that they arise from pairwise, antagonistic combinations of the outputs from two classes of cone photopigments (Michael, 1968; Tong, 1977; Jacobs & Tootell, 1980). Indeed, in a previous chapter, an illustration was presented to suggest that there is a close fit between the wavelength-discrimination properties of such cells and comparably obtained behavioral data (see Figure 3.24). On the other hand, however, it has also been reported that these chromatically opponent responses are not representative of a physiologically normal visual system, but that rather they represent artifacts induced by particular experimental preparations (Gur & Purple, 1978). Even though there is in turn some evidence against this claim (Jacobs & Tootell, 1980), it cannot yet be fairly concluded that this issue is settled.

In the past few years the results from behavioral studies of color vision have been reported for several species from among the ground-dwelling sciurids. In one series of investigations, visual sensitivity and color vision have been measured in Mexican and thirteen-lined ground squirrels (Jacobs and Yolton, 1971), the California ground squirrel (Anderson & Jacobs, 1972), the prairie dog (*Cynomys ludovicanus*) (Jacobs & Pulliam, 1973), and the golden-mantled ground squirrel (*S. lateralis*) (Jacobs, 1978).

In each of these studies animals were tested in a three-alternative, forced-choice discrimination task. After a determination of their spectral sensitivity functions, the animals were further tested on various color discrimination problems using luminance equations based on the sensitivity measurements and on direct brightness matches. The basic test involved a discrimination between equiluminant monochromatic and achromatic lights. The outcome of this test for all five species is summarized in Figure 5.1. From the figure it is apparent that all of these animals were successful in discriminating various chromatic lights from equally luminant achromatic lights. This means that all of these squirrels possess color vision. However, discrimination performance falls to chance levels (33% correct in this case) at a single spectral location. This indicates the presence of a spectral neutral point and thus the color vision of these ground-dwelling sciurids can be formally classified as dichromatic. Also worthy of note in Figure 5.1 is the narrowness of the region of neutrality (much as it is for human dichromats), and that all five species have similar color vision in that the neutral-point locations are closely similar for all animals. Several other measures of color vision summarized in this same paper (Jacobs, 1978) verifies the basic dichromatic nature of color vision in these animals.

In addition to the five species of ground-dwelling sciurids whose data are summarized in Figure 5.1, there has also been a study of the antelope ground squirrel (*Ammospermophilus leucurus*) by Crescitelli and Pollack (1972). These investigators tested animals in a two-alternative discrimination task. Potential brightness cues were eliminated, or made irrelevant, by using sensitivity mea-

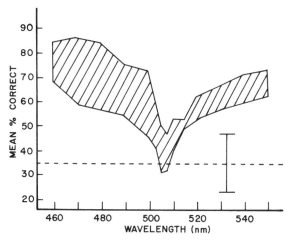

Figure 5.1   Color discrimination by ground-dwelling sciurids. The hatched area encloses the performance levels (in mean % correct) of sciurids tested on discriminations between various monochromatic and equiluminant white lights. The results are for 14 animals drawn from 5 different species. Chance performance was 33% (horizontal dashed line). The vertical bar shows the 95% confidence interval. Note that these animals were able to successfully discriminate all wavelengths except those in a narrow spectral band centered at 505 nm. This region defines the location of the neutral point in these dichromatic animals. (After Jacobs, 1978.)

surements based on the electroretinogram, and also by wide variations in stimulus luminance. The antelope ground squirrels were able to successfully discriminate between several pairs of opposed stimuli; for example, a 460-nm light versus "white," and 460-nm light versus other spectral wavelengths located at 500 nm and longer. However, some stimulus combinations were not successfully discriminated, leading Crescitelli and Pollack to conclude that this species has dichromatic color vision. If so, it appears that the dichromacy of this species may be quite different from that shown by other ground squirrels because the antelope ground squirrels were unable to successfully discriminate a 520-nm light from several other stimuli. In fact, the authors suggest that for the antelope ground squirrel, 520-nm lights are "seen as achromatic" (contrast this result with that shown in Figure 5.1). Thus, although it seems very likely that several species of ground-dwelling sciurids do have dichromatic color vision, it is not yet certain if the nature of this dichromacy is identical for all squirrels.

The arboreal cousins of the ground squirrels, the tree squirrels, have also provided attractive targets for color vision researchers. An examination of the eyes of these animals suggested to Walls (1942) that they have an all-cone retina. This contention turned out to be incorrect. The photoreceptors in the tree squirrel retina are in fact arranged into two distinctive tiers (see Figure 3.11), and several investigators (Cohen, 1964; West & Dowling, 1975; Anderson & Fisher, 1976) have used a variety of different techniques to examine the anatomy of the two types of receptors. Their unanimous conclusion is that the inner-tier receptors, those located closer to the vitreous, are rods, whereas the outer-tier receptors are cones. Rods are considerably more common in the tree squirrels than in ground squirrels, but even in tree squirrels the total number of rods is smaller than the total number of cones (Cohen, 1964; Yolton, 1975).

As in the case of the ground squirrel, attempts to characterize the photopigments in the tree squirrel retina have been less than completely successful. In 1955, Weale used the newly developed technique of reflection densitometry to examine the gray squirrel (*Sciurus carolinesis*) retina. He found evidence for a single photopigment having a 530-540-nm peak, although the sensitivity curve for this pigment was unusually narrow relative to those generally found to characterize photopigments. Later, Dartnall (1960) was able to extract what was apparently a rod photopigment (peak at 502 nm) from the gray squirrel retina. And still more recently, microspectrophotometry (MSP) has been applied to an examination of single photoreceptors in the gray squirrel eye (Loew, 1975). In this investigation those photoreceptors located in the inner tier (rods) were found to show maximal absorbance at 500 nm whereas those photoreceptors making up the outer tier (cones) showed an absorbance peak at 540 to 545 nm. No other photopigments were found in any of the outer tier receptors in a sample of more than 70 such cells. This latter fact is significant because there is other evidence (see the following) that there are probably two classes of cone photopigments in the tree squirrel retina.

The color vision of fox squirrels (*Sciurus niger*) (Jacobs, 1974) and Western gray squirrels (*S. griseus*) (Yolton, 1975) has been examined behaviorally in forced-choice discrimination tasks. Both of these species were found to be able to discriminate a wide range of spectral lights from equiluminant achromatic lights. Although no completely compelling evidence for the presence of a neutral point was found in these experiments, further measurements of wavelength discrimination (Jacobs, 1976) suggest that these squirrels are in fact dichromatic. The results of this latter experiment are shown in Figure 5.2. Wavelength discrimination thresholds for two Western gray squirrels are plotted in this figure. These animals showed best discrimination in the vicinity of 500 nm (where they were able to discriminate differences of as little as 3 nm). Discrimination ability fell off sharply for test wavelengths both longer and shorter than 500 nm, and those animals were completely unable to discriminate 550 nm from 604 nm. Wavelength discrimination functions of this type are very similar to those usually obtained from human dichromats (see Figure 2.16). In support of this conclusion, Silver (1975) has also found wavelength-discrimination behavior consistent with a diagnosis of dichromatic color vision for the gray squirrel.

To summarize, work done thus far on the sciurids suggests that at least some

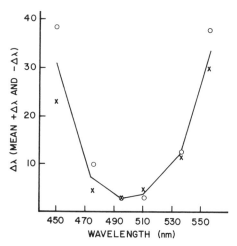

Figure 5.2 Wavelength discrimination in the gray squirrel. The symbols represent the performance levels achieved by two animals. (After Jacobs, 1976.)

features of vision are much the same for both ground squirrels and tree squirrels. Thus, both have rods containing a 500-nm photopigment, although those receptors are much more robustly represented in tree squirrels. In addition, both types of squirrels seem to have two types of cone photopigments that provide the basis for dichromatic color vision. In both cases, the retinas lack the long-wavelength photopigment characteristic of the normal human retina, and so color vision in both groups of animals appears roughly similar to that of human protanopic dichromats. Whether the dichromacy of these two types of squirrels is exactly the same remains unknown.

B. Rat

In many ways the rat seems one of the most unlikely subjects for a color vision experiment. Of course, rats are strongly nocturnal, but, beyond that, it is obvious from even casual observations that these animals rely very heavily on nonvisual information in the conduct of their normal behavior. Nevertheless, due no doubt to their status as a standard laboratory subject, there is a long, controversy-ridden history of research on the vision of the rat. The very presence of cones in the retina of this rodent has sometimes been doubted. Lashley (1932), for instance, claimed the rat retina was composed of only rods, although only 2 years later Walls (1934) published a picture of a cone from the retina of a Norway rat and noted that cone-type nuclei were ''abundant'' in the rat retina. A strong claim for rat cones was made by Sidman (1958) who argued that cones in the rat retina are similar in appearance to the cones found in the retinas of other mammals. In fact, he reported that cones make up a rather surprising 10–20% of the total number of photoreceptors in the rat. Others have also found some manifestations of cones in the rat retina (Ladman, 1958; Dowling, 1967; LaVail, 1976). It now seems clear that this rodent does have an anatomically duplex retina.

Evidence that rods and cones provide for functioning scotopic and photopic systems in the rat comes from a number of studies, both behavioral and physiological in approach. Although there is some argument about the mechanism, and about the magnitude of the effect, these studies, taken as a group, show that rat spectral sensitivity changes as a function of the adaptation state of the eye. Beyond that, there are two direct claims about the photopic mechanisms in the rat retina. In behavioral experiments involving the use of chromatic adaptation, Birch and Jacobs (1975) found evidence for at least two such mechanisms having sensitivity peaks at about 520 and 540 nm, while Cicerone (1976) used the electroretinogram (ERG) and intense chromatic adaptation to obtain sensitivity curves suggesting the presence of three photopic mechanisms with peaks at about 450, 520, and 560 nm. Finally, in the course of recording from the lateral geniculate nucleus of the albino rat, Lemmon and Anderson (1979) encountered one cell that showed spectrally opponent response properties. Further details about the characteristics and incidence of such units are not yet available.

Although there was an early claim that rats have no color vision (Watson & Watson, 1913), the experiment presented in its support was clearly inadequate to answer the question. During the 1930s a number of psychologists turned their attention to the question of rat color vision. The result was a series of research reports that included claims both pro and con as to the presence of color vision in this species. Probably the two most widely cited studies reached quite opposite conclusions. Ironically, the two studies employed roughly similar methods, and both are fairly convincing in their claim.

The first study was done by Coleman and Hamilton (1933). They trained hooded rats in a Lashley jumping stand to leap toward one or the other of two panels. These panels were painted, in varying shades, either gray, red, blue, or green. The animals were initially trained to jump toward the darker of a stimulus pair. Once this was learned, the relative lightnesses of the panels were systematically changed until the brightness discrimination broke down. The pair of stimuli where the discrimination failed, a so-called "confusion pair," defined a brightness match between the two stimuli. Thereafter, the experimenters tried to establish a successful discrimination between these confusion pairs. Their efforts failed completely, both with the initial group of rats and with fresh animals that had not received any initial training on a brightness discrimination. Reasonably, Coleman and Hamilton concluded that the rat is color-blind.

Toward the end of the same decade, after a number of other positive claims for rat color vision had been reported, Walton and Bornemeier (1939) published a second widely cited study. They tested both albino and pigmented rats in a jumping stand task. The stimuli were spectral lights, obtained by placing gelatin filters in front of tungsten-filament lamps. Like Coleman and Hamilton, they obtained luminance equations by initially training animals in heterochromatic brightness discrimination tasks and then reducing the luminance difference between a stimulus pair until an equation was obtained. Following this, new animals were tested on several pairs of stimuli: red versus blue, red versus green, blue versus yellow, blue versus green, and green versus yellow. The animals were able to successfully discriminate between all but the latter two of these combinations. To assure that a luminance mismatch was not responsible for the successful discriminations, they did some additional experiments that involved programmed variations in luminance about the equation points. Again, the animals were able to show successful discrimination. Walton and Bornemeier concluded that the rat does have color vision. Obviously, as the authors pointed out, this capacity is not a very acute one—all of the successful discriminations involved pairs of stimuli having dominant wavelengths that differed by more than 50 nm. Nevertheless, this experiment appears impeccable, and its evidence for the presence of color vision in the rat is compelling.

There is no very satisfactory way of reconciling these two experiments. However, when a sensory capacity is a feeble one, as color vision so clearly is in the rat, it is probably reasonable to put greater weight on positive than on negative results (on the grounds that it is more difficult to detect a weak signal than it is to fail to detect it). This seems even more reasonable here because of the other evidence for photopic vision in the rat. Nevertheless, it would probably be worthwhile to have rat color vision reinvestigated with modern techniques. For one thing, the tests should be carried out at light levels sufficiently high to rule out the intrusion of signals from rods. It is not certain if this was the case in the early studies. Beyond verifying the presence of color vision, which seems likely, it would be useful to learn something of the nature of color vision in a species that appears to have retained (or possibly acquired) only a minimum of a photopic capacity. One wonders if rat color vision might be a rod-diluted version of the color vision found in the strongly diurnal rodents, such as the ground squirrels.

The research on rat color vision may have been worthwhile in a more general sense. First, it provided clear examples of the need for careful control of possible brightness cues in tests of animal color vision. In doing so it sensitized many subsequent experimenters to one of the major problems of research in this area. Second, if one can extrapolate from work on the rat, it seems plausible to expect that even the most resolutely nocturnal mammals will be found to have at least a rudimentary color vision capacity.

II. The Domestic Cat

Because of its nocturnality, the cat represents another species that would logically seem an unlikely target for color vision experiments. The reason for interest in cat color vision derives in large part from its popularity as a subject for study by the visual physiologists. In particular, the pioneering experiments of Granit suggested to many that the physiological machinery necessary to support color vision was present in the cat retina, although Granit (1955) himself cautioned that a putative retinal mechanism for color vision did not necessarily imply the presence of this behavioral capacity. Some have suggested that interest in cat color vision has also been fueled by its status as a common house pet (Daw, 1973). If this is so, it makes it hard to understand the virtual absence of good laboratory research on the color vision of an equally popular companion, the dog (see page 157). At any rate, there have by now been a fair number of experimental studies relevant to the question of color vision in the cat.

As expected of a primarily nocturnal animal, the retina of the cat is very rod rich. However, there are also substantial numbers of cones in this retina. For example, cone density increases dramatically in the region subserving central vision in the cat eye, the *area centralis,* reaching something on the order of 25,000 cones per mm^2 (Steinberg, Reid, & Lacy, 1973). Although this figure is about six times lower than that for a comparable location in the human retina, it is clear that the photoreceptor content of the cat retina implies duplex vision.

There are no direct measurements of the retinal photopigments in the cat retina. A variety of other indices, both electrophysiological and behavioral, show that the rod photopigment is like that in other mammalian retinas having peak sensitivity at about 500 nm. There are a few estimates of the cone photopigment spectra obtained from electrophysiological recording experiments. One suggests that the cat retina contains two classes of cone photopigments having peak sensitivities at 445 and 555 nm (Pearlman & Daw, 1970), whereas another claims evidence for three photopigment classes having peaks at 450, 500, and 555 nm (Ringo, Wolbarsht, Wagner, Crocker, & Amthor, 1977). Saunders (1977) has additionally claimed the presence of what would be for a mammal an exceptionally long-wavelength photopigment ($\lambda_{max} \approx 600$ nm). Based as they are on scanty data and various extrapolations, all of these figures should be considered as quite tentative.

Some brief comments on possible neural bases for cat color vision can be made. On occasion single cells have been found in the retina and lateral geniculate nucleus that show typical spectral opponency between underlying components having maximal sensitivity at about 450 and 565 nm. These units appear to be very rare, or at least are not easily recorded. For example, Pearlman and Daw (1970) found only three such cells out of a total sample of 118 cells. The paucity of chromatically opponent cells has led Hammond (1978) to suggest that at light levels below which the rod signals have become saturated, cat color vision is likely based on interactions between the outputs from rods and the 565-nm cones.

A rather different picture emerges from a single-cell recording study carried out by Ringo *et al.* (1977). These investigators found evidence for three spectral mechanisms active at photopic light levels with peaks at 450, 500, and 555 nm. Most novel is their claim that a 500-nm mechanism resides in cat cones as well as in cat rods. They note that although the inputs to ganglion cells from these three mechanisms are typically added together, they also found some ganglion cells in which ''the various inputs were antagonistic.'' The discovery of three photopic mechanisms leads them to the prediction that the cat may eventually be found to have trichromatic color vision (Ringo *et al.,* 1977).

As in the case of the rat, behavioral work on cat color vision has also not been free of contradiction. Based on natural history considerations, early writers often assumed that the cat lacked color vision. Walls (1942), a scientist refreshingly free of equivocation, concluded that, ''we can be quite sure that the cat has no hue-discriminating capacity at all.'' Of the modern studies on cat color vision, the weight of evidence shows Walls' view to be clearly incorrect. Although some experimenters could find no evidence for color vision in the cat (Meyer, Miles, & Ratoosh, 1954; Gunter, 1954), all of the more recent studies on the subject report positive results (Mello & Peterson, 1964; Clayton & Kamback, 1966; Daw & Pearlman, 1969; Brown, Shively, LaMotte, & Sechzer, 1973; Loop & Bruce, 1978). The reasons for the differences in outcomes are not completely clear and probably not very informative to pursue. For one thing, it is clear that cat color vision experiments require a good deal of experimenter perseverance because this animal appears to be much more predisposed to operate on cues other than color. Extended periods of training are required to demonstrate color vision in the cat. In this regard the cat is like other species in which color vision is not highly developed or, possibly, does not provide a very compelling source of environmental information.

One interesting possibility as to why it may have been difficult to demonstrate color vision in the cat is suggested by the work of Loop and Bruce (1978). They discovered that although it was relatively easy to train cats to discriminate colors when the test stimuli used were very large in area, it became progressively more difficult to find these same successful discriminations when the stimuli were decreased in size. Indeed, when the stimuli were reduced to smaller than about 20° of visual angle, the cats were quite unable to make any color discriminations. Whether this fact can be used to resolve all of the earlier conflicting results from studies of cat color vision is uncertain. Nevertheless, it does suggest that stimulus size may be an important parameter to consider in studies of color vision in species that, like the cat, have relatively low densities of retinal cones.

Although the cat clearly possesses some color vision, the nature of this capacity is not yet unambiguously defined. In the most searching investigation to date, Brown *et al.* (1973) trained cats in a two-choice discrimination apparatus. After first carefully matching four spectral stimuli (roughly red, yellow, green, and blue by human hue designations) to be equally bright for the cat, they then attempted to train six animals to discriminate between all pairwise combinations of these four stimuli. Although substantial training was required, the animals eventually learned to discriminate between each of these pairs except for the pairing of yellow versus green. These two stimuli were indiscriminable, even with extended training and with the use of specialized tutoring procedures. This result poses a problem because, if the cat has two cone pigments peaking at roughly 445 and 555 nm, or three cone pigments peaking at 450, 500, and 555 nm, as previous estimates suggest, then it is difficult to see how the cat could successfully discriminate between red and yellow stimuli but fail to discriminate the yellow versus green combination. Furthermore, Brown *et al.* (1973) argue that the entire middle range of the spectrum, from yellow through green, is indistinguishable from white for the cat. This would of course be quite unlike the neutral points found in those other color vision systems based on two photopigment classes. One can only conclude that not all is yet known about color vision in the cat.

III. Nonhuman Primates

There has never been any argument about the widespread presence of color vision among the primates. Indeed, there are repeated references in the general literature on primates to the importance of color in the conduct of primate behavior. Much of this evidence is anecdotal, and it is sometimes rather dramatic. For example, consider the following revelation:

> Monkeys are extremely colour conscious. One owned by the author would make violent love to a young lady clad in a scarlet coat, and permit her to nurse and fondle him. Later in the day he would attempt to scalp her, amid screams of rage, when she again appeared dressed in light blue [Smythe, 1961].

The experimental literature on color vision in the nonhuman primates is much more restricted than one might imagine. Large numbers of species are still entirely uninvestigated. This research area was the subject of a comprehensive literature review written several years ago (De Valois & Jacobs, 1971). Consequently, I shall summarize this research in a somewhat selective manner, emphasizing only those studies that are the most informative. For a more complete list of references, particularly to the earlier studies in this area, the reader is referred to De Valois and Jacobs (1971).

A. Apes

In the early 1940s, Grether (1940a,b,c,d, 1941) conducted an extensive series of studies on color vision in the chimpanzee (*Pan*), measuring wavelength and colorimetric purity discrimination as well as comparing the wavelength limits of the visible spectrum for man and chimp. These experiments are notable ones, not only because they involved the use of an appropriate test apparatus and good experimental technique, but also because Grether consistently tested human subjects and chimps in the same situations so as to provide useful comparative data. To measure color vision, Grether trained chimps in a two-choice discrimination apparatus in which the stimuli (various achromatic and monochromatic lights) were projected onto white discs. The animals were trained to reach toward one of these discs in order to gain access to a food reward.

Grether (1940a) concluded that the wavelength-discrimination abilities of chimp and man were highly similar because the wavelength-discrimination functions for the two species had the same general shape. The absolute levels of discrimination were also closely similar, except in the very long wavelengths where the difference thresholds were clearly higher for chimp than for man. This comparison is shown in Figure 5.3. Despite the difference in discrimination capacity in the long wavelengths, the overall form of the function suggests similar color vision for these two primate species. Grether (1941) found that performance in a colorimetric-purity discrimination task was also much the same

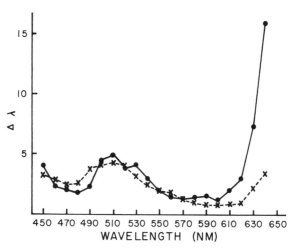

Figure 5.3 A comparison of wavelength discrimination by chimpanzee (●—●) and man (×- - -×). The chimpanzee data are means for three animals. The comparison curve was obtained from a single human subject. All subjects were tested in the same two-choice discrimination apparatus. (Data taken from Grether, 1940a.)

for man and chimp. Both appeared to perceive the spectral extremes as highly saturated, and both showed a clear saturation minimum in the yellow region of the spectrum. The location of this minimum was very slightly different for the two species, being located at 575 nm for the human and at 570 nm for the chimp.

In still other experiments, Grether (1940b) examined color mixing by the chimpanzee. Two such mixtures were examined. One was a Rayleigh-type match in which the relative proportions of red (640nm) and yellowish-green (560 nm) lights needed to match the appearance of a yellow light were determined. The required mixture proportions were found to be almost precisely the same for chimp and man. One can conclude from this that chimpanzees have neither a protan nor a deutan color deficiency. In a second color-mixing experiment, the proportions of 610- and 495-nm lights necessary to match an achromatic light were measured. In this experiment the mixture proportions required were somewhat different for the two species with the chimpanzees needing a slightly higher proportion of the 610-nm light than did the human subjects. Because the Rayleigh match equation implied a virtual identity of the color vision systems subserving green through red for man and chimp, the outcome of this latter experiment might be explained as resulting from a slightly higher sensitivity of the chimps to the short wavelengths, perhaps due to differences in the degree of preretinal absorbance in the two species (De Valois & Jacobs, 1971). At any rate, the differences between the color vision of man and chimpanzee in these several experiments were either small or nonexistent. It seems reasonable to conclude, as Grether did, that the modal color vision of man and chimpanzee is highly similar, if not completely identical.

Other than Grether's studies of the chimpanzee, very little is actually known about ape color vision. In one study, Tigges (1963) did obtain some evidence for color vision in three gibbons (*Hylobates lar*) and one orangutan (*Pongo*). Unfortunately, this study does little more than verify the presence of color vision in these species. Although it is probably not unreasonable to assume that all of the great apes have color vision similar in character to that of man, it would be well to have this verified experimentally. Considering the large number of captive apes who spend substantial portions of their lives staring out of small enclosures, it would probably provide some diversion to them, as well as a benefit to us, to engage their cooperation and intelligence in a range of experimental ventures, including a detailed determination of the characteristics of their color vision.

An additional point about ape color vision deserves mention. The large expanse of neocortex characteristic of the ape brain strongly suggests that the color vision of these species might be studied in ways that it probably cannot be in other animals. Evidence justifying this belief has been available for some time. For example, Kohts (1928) noted that color seemed to be a very important stimulus dimension for the chimpanzee. She found that the chimp was easily able to match the colors of objects having quite different shapes. Perhaps even more impressive, on one occasion Kohts presented a chimpanzee with a large collection of colored chips and found that without any instruction whatsoever the animal proceeded to sort out these chips according to their differences in color!

Over the past several years a number of research projects have attempted to show that language systems can be acquired and creatively employed by various ape subjects. Avoiding the highly controversial question of whether these projects actually demonstrate the occurrence of a true language, it is at least possible that with a means of interspecies communication a way to access their color experiences directly is now available. One such attempt has already been reported. A group of investigators (this work is summarized by Rumbaugh, 1977) have taught a young female chimpanzee, named Lana, a visual language in which the ape learned to employ a large number of lexigrams to represent language components. Among the "words" available in this language were eight color names—white, black, blue, orange, green, purple, red, and yellow. In several test situations Lana was consistently able to successfully name the color of one indicated object from among several objects that were presented simultaneously. Beyond that, Lana was also tested in a situation where she was required to supply color names for an extensive collection of standard Munsell color chips varying in hue, lightness, and saturation. Like Koht's subject, Lana also gave evidence that color is an important stimulus dimension for the chimpanzee. Indeed, the investigators noted that Lana appeared to find her color-naming chores "very enjoyable." The result of this naming experiment was that the chimpanzee used color names in much the same way as a human with normal color vision. That is, the locations of the spectral boundaries between hues seemed much the same for man and chimpanzee. The possibilities for further direct investigation of chimpanzee color experience appear substantial and intriguing.

B. Old World Monkeys

Various species from the genus *Macaca* have served as subjects in many vision experiments, including a fair number directed toward color vision. The popularity of these monkeys rests principally on the commonplace belief that their visual systems are highly similar to that of man so they can serve as reasonable surrogates in experiments that would be impossible with human subjects. With regard to color vision, the evidence that man and macaque monkeys have identical capacities will be considered shortly. Before doing so, two related points need to be made. First, it should be noted that there have been a large number of studies of the electrophysiology of the macaque visual system. Many of these involved analyses of the responses of single cells to spectral stimuli, and thus they provide a matrix for understanding the mechanisms for color vision in these monkeys. Some of these results were summarized in Chapter 3 and references to these studies can be conveniently found in several review articles cited there (especially Abramov, 1970; Daw, 1973; De Valois, 1973; De Valois & Jacobs, 1981).

A second point of comparison between the visual systems of man and the macaque monkey is that recent measurements show the photopigments of these species to be closely similar. Using MSP, Bowmaker and Dartnall (1980) found the three classes of cone pigments in man to have λ_{max} of 420, 534, and 567 nm. The comparable figures for the cynomolgus macaque (*M. fascicularis*) were 415, 535, and 567 nm (Bowmaker *et al.*, 1980). For the rhesus monkey (*M. mulatta*) cone pigments having λ_{max} at 536 and 565 nm were found (Bowmaker *et al.*, 1978). For this latter species no cones have yet been found to contain a short-wavelength photopigment. However, its absence in these measurements should quite certainly not be taken to indicate that it does not exist. The important result from these MSP measurements is that the cone pigments of two common species of macaque monkeys and of man appear to be virtually identical.

It was noted earlier that there is some evidence for significant variations in the absorbance peaks of photopigments found in the retinas of individual animals of the same species. Bowmaker *et al.* (1980) report a similar finding in the primate retina. For instance, measurements made on 14 different cones belonging to the 567-nm class in the retina of the cynomolgus monkey had individual peaks covering the range from 554 to 575 nm. Because repeat measurements made on the same cone did not differ by more than 4 nm, there is clearly a real and substantial variability in the absorbance characteristics of photopigments nominally placed into the same class. Further, not all of the variability was due to differences between animals as a range of different λ_{max} values was also noted within individual retinas (Bowmaker *et al.*, 1980).

Following a number of early studies, most of which were able to provide little more than the conclusion that macaques do have color vision, Grether (1939) conducted the first detailed study of these animals. He tested five rhesus monkeys in essentially the same two-choice discrimination apparatus as that in which the chimp studies described previously were conducted. Three different tests of color vision in the rhesus monkey were performed. First, Grether showed that the rhesus monkey easily discriminates spectral wavelengths between 480 and 600 nm from what were (apparently) equiluminant white lights. This implies both that color vision is present and that the color vision is not of the dichromatic variety because no neutral points were discovered. Second, Grether measured wavelength discrimination at three locations: 500, 589, and 640 nm. These results, along with comparison measurements for several other species, are given in Table 5.1. At each point the difference thresholds for the rhesus monkeys were not significantly different from those obtained from normal human trichromats tested in the same apparatus.

Finally, the relative proportions of a mixture containing 495- and 610-nm

Table 5.1

Wavelength Discrimination Thresholds for Several Primate Species as Measured at Three Different Spectral Locations[a]

Species	Number tested	Test wavelengths[b]		
		500 nm	589 nm	640 nm
Humans	3	12	1	12
Rhesus monkeys	5	9	3	11
Pig-tailed monkeys	1	7	4	—
Green monkeys	1	—	4	—
Baboons	1	9	2	11
Cebus monkeys	3	8	6	38
Spider monkeys	1	10	2	6

[a] From Grether (1939). All subjects were tested on the same two-choice discrimination apparatus.

[b] For the 500-nm test wavelength, discrimination was measured toward the longer wavelengths while for the other two values discrimination was measured toward the shorter wavelengths.

lights that were required to match the appearance of an achromatic light were determined. In this case the rhesus monkeys required somewhat more red light (or less blue light) in the mixture than did the human subjects. Grether did not interpret this as implying color-defective vision in the rhesus monkey although, perhaps paradoxically, he did claim this difference to be a statistically significant one.

Thirty-five years after the Grether experiments, De Valois, Morgan, Polson, Mead, and Hull (1974) reported results from another series of experiments carried out on macaque monkeys. These investigators utilized a four-choice discrimination apparatus similar in concept to that illustrated in Figure 2.8. They tested representatives from three species of macaques: cynomolgus, pigtail (*M. nemestrina*), and stumptail (*M. speciosa*) monkeys. After first running several experiments on visual sensitivity, and thus providing the groundwork necessary for making luminance equations, they carried out four experiments on color vision: neutral-point tests, Rayleigh match tests, wavelength discrimination, and colorimetric purity discrimination. The results from these experiments can be summarized very economically: in all tests the color vision capacities of the macaque monkeys were qualitatively indistinguishable from those measured on normal human trichromats tested in the same situation. Furthermore, there was remarkably little difference between the species along any quantitative dimension. For example, Figure 5.4 shows the wavelength-discrimination data obtained from three pigtail and two cynomolgus macaques along with some com-

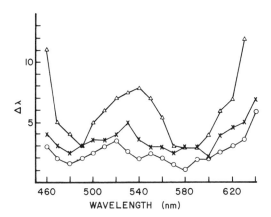

Figure 5.4 Wavelength discrimination functions for three primate species tested under equivalent conditions. *Macaca nemestriana* (×); *Macaca fascicularis* (△); human (○). (Data from De Valois *et al.*, 1974.)

parative data obtained from humans tested in the same apparatus. The curves from all three species show the familiar normal trichromatic form (see Figure 2.11) with twin minima located at about 490 and 590 nm. On an absolute basis the thresholds for the monkeys were somewhat higher than those for the humans. These thresholds are also perhaps slightly different for the two macaque species, although the difference is not great, and the investigators did not believe it to be a significant one.

Recently, Oyama, Furusaka, and Kito (1979) measured Rayleigh matches in both rhesus and Japanese (*M. fuscata*) macaques. The result was that these species performed more like normal human trichromats in terms of the red–green proportions required to match a standard yellow light than either protanomalous or deuteranomalous humans tested in the same situation. In sum, five different species of macaque monkeys have been examined in three independent sets of experiments. In each case the conclusion has been that their color vision is most similar to that shown by normal human trichromats who were tested in an equivalent way.

Despite this concordance of opinion about macaque color vision, in a recent investigation the question has been raised as to whether the color vision of the rhesus monkey as measured with foveal stimuli might differ significantly from normal trichromacy. To reach this conclusion, Zwick and Robbins (1978) measured spectral sensitivity functions using a standard acuity target, the Landolt ring, as a test stimulus. When a coarse acuity criterion was employed (that is, a large gap in the ring) the spectral sensitivity functions obtained from rhesus monkeys and normal human trichromats were not systematically different. However, when a fine acuity criterion was employed, the rhesus monkeys had significantly lower sensitivity at the long wavelengths (around 600 nm). In fact, in the latter case the spectral sensitivity functions for the rhesus monkeys fell off more in the long wavelengths than did a function obtained from a human protanomalous trichromat. Zwick and Robbins (1978) interpret these results to suggest that the rhesus monkey may have protanomalous color vision if the stimuli are restricted to the region of the fovea. This experiment did not involve any direct measurement of color vision, but given that a loss of long-wavelength sensitivity is a defining characteristic of protanomalous color vision, then it may be that the mechanisms underlying color vision in the rhesus foveal projection are significantly different from those available extrafoveally. In this view, the normal trichromacy claimed for the macaque in other experiments would reflect the use of large stimuli that do not exclusively engage foveal vision.

Physiological evidence supporting this possibility has recently been reported by De Monasterio and Schein (1980). They found that ganglion cells in the macaque monkey that respond in a spectrally nonopponent fashion differ systematically in their spectral sensitivities depending on their retinal locations. Specifically, cells having inputs from the fovea showed lower sensitivity to long wavelengths than did cells having more peripherally located inputs. This variation is consistent with the behavioral results just described. Despite both behavioral and physiological evidence for regional sensitivity variations in the macaque retina that do not coordinate with that characteristic of the human retina, it is important to emphasize that there is so far no evidence to indicate similar variations in color vision. Nevertheless, it is an issue that merits further study.

In view of all of the similarities between the color vision of man and macaque monkey, it is not surprising that the hue categories employed by these species also appear similar. This fact was established in an experiment in which monkeys were trained to press a key in the presence of a chromatic stimulus (a training wavelength) in order to receive a liquid reinforcement (Sandell, Gross, & Bornstein, 1979). After acquisition of this behavior, the animals were tested to see how long they would continue to respond in the presence of a new chromatic stimulus (a test wavelength) even though the response no longer produced reinforcement. The essential finding was that the monkeys responded more persistently in the presence of test wavelengths that fell within the same human hue category (blue, green, yellow, etc.) as the training wavelength than they did if the training and test wavelengths had not been drawn from the same hue categories irrespective of the actual wavelength differences between training and testing wavelengths. The conclusion is that humans and macaque monkeys sort the spectrum into the same hue categories.

In 1971, Humphrey carried out a very novel experiment that attests to the importance of color vision to macaque monkeys. He was interested in assessing the preferences shown by monkeys for various stimulus dimensions including brightness, hue, and spatial content. The test situation was such that the monkey indicated (by pressing a lever) which of two equally available stimulus displays it preferred to look at when there was no differential reinforcement associated with selecting either one. Humphrey found that when the animals were tested with various pairs of stimuli, one spectral and one an equiluminant white, they consistently chose one or the other of these stimuli. Relative to white light, the monkeys showed positive preferences for blue and green stimuli, and increasingly strong negative responses for yellow, orange, and red lights. These results are illustrated in Figure 5.5. The fact that the animals showed such a consistency of choice among equiluminant stimuli indicates once again that they have color vision. Much more interesting than the reestablishment of a fact already well-known was the discovery that the nature of these preferences was highly consistent among animals and that, perhaps even more startling, these monkeys showed preferences for colored versus achromatic lights in proportion to their wavelength content; that is, the degree of preference decreased as the wavelength of the light was increased! The reasons for this particular ordering of colors, indeed for why color preference should appear contant across members of the species, remain unknown although Humphrey (1972) has advanced some speculations on the matter.

   To summarize, the bulk of measurements made on several species of macaques strongly indicate that their modal color vision is greatly similar to, if not identical with, that of normal human trichromats. This obviously makes these species highly useful as subjects in studies bearing on the mechanisms underlying color vision.

   Other than the results for macaque monkeys, little is known about color vision in any of the other Old World monkeys. Grether (1939) himself carried out a few tests on a baboon (*Papio papio*) and on an African green monkey (*Cercopithecus sabaeus*). Insofar as they were tested, these animals did not appear to be greatly different from the macaques (see Table 5.1). Beyond Grether's brief observations, the literature on color vision in other Old World monkeys is either nonexistent or exceedingly well concealed. Thus, although it is commonly assumed that

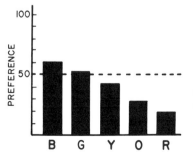

Figure 5.5 The preferences shown by rhesus monkeys for various colored lights (blue, green, yellow, orange, and red) versus that shown for equiluminant white lights. Values greater than 50 indicate a positive preference for the colored light; values less than 50 indicate a positive preference for the white light. (Data taken from Humphrey, 1971.)

all of the Old World monkeys have color vision similar to that of the macaques (and man), this assumption has a vanishingly small supporting base.

C. New World Monkeys

Interest in the New World monkeys has been stimulated, at least to some extent, by the possibility that color vision of the members of this group may very well have evolved along pathways quite separate from those followed by the Old World monkeys. Because of this, as well as for other reasons, various theorists (for example, Clark, 1947) have at times speculated that one might find among the New World monkeys forms of color vision analogous to those characteristic of various types of human color defects; for example, an entire species might have dichromatic color vision. Unfortunately, as we shall see, at the present the only completely certain general conclusion about color vision among the New World monkeys is that no *species* has been shown to completely lack the capacity, nor has any *species* been conclusively shown to have normal trichromatic color vision.

The New World species that has perhaps received the most attention is the *Cebus* monkey. In his comparative studies of monkey color vision, Grether (1939) tested three *Cebus* monkeys (two *Cebus capucinus* and one *Cebus unicolor*). The test apparatus and the procedures were the same as those used with macaque subjects, and so too, apparently, were the luminance corrections, which to some unknown extent might have created a problem. The wavelength-discrimination capacities shown by these three animals were significantly poorer for long-wavelength stimuli than those of the Old World primates. Thus, as indicated in Table 5.1, the difference thresholds measured at 640 nm were about three times the size shown by color-normal observers. Furthermore, in a color mixture test the *Cebus* monkeys required substantially greater amounts of 610-nm light in mixture with 489-nm light in order to match an achromatic standard. One *Cebus* monkey was also tested for the possible presence of a spectral neutral point. Over the range from 600 to 480 nm, examined in 10-nm steps, the animal was able to successfully discriminate spectral lights from white except when the spectral light was set to either 520 or 510 nm. At these two wavelengths the animal failed to make the discrimination. Grether concluded from all of these results that the *Cebus* monkey has dichromatic color vision, probably protanopia.

Several years later, Malmo and Grether (1947) reported further measurements on an additional *Cebus* monkey. These results confirmed the earlier study in that this animal was also unable to successfully discriminate 520- and 510-nm light from an achromatic standard. Again, the conclusion was that the *Cebus* monkey is a dichromat.

Two further studies have raised serious questions about this conclusion. In one study, Gunter, Feigenson, and Blakeslee (1965) tested hue discrimination in five *Cebus* monkeys. Although this study is somewhat more difficult to interpret because the experimenters used colored papers as stimuli rather than monochromatic lights, they did find that the monkeys were capable of making some successful discriminations among long-wavelength stimuli that would appear to be impossible for a dichromatic observer. Furthermore, they also found that their subjects were able to discriminate between blue–green stimuli and equally bright gray ones. This latter result also appears contrary to the Grether results.

De Valois and Morgan (personal communication, 1976) have also examined several aspects of visual sensitivity and color vision in *Cebus* monkeys. Their general conclusion is that, although this species does not have normal trichromatic color vision, the defect is less severe than the dichromacy that Grether believed to characterize these animals. Specifically, they have found that, first, these animals do show some loss in spectral sensitivity for long test wavelengths relative to normal trichromatic observers. Second, in a thorough examination of various spectral locations, they have also demonstrated that the *Cebus* monkey is able to discriminate all monochromatic lights from equiluminant achromatic ones, including those from the band around 500 nm, which both Grether (1939) and Malmo and Grether (1947) found their animals unable to discriminate. It seems likely that this difference in outcome between studies reflects training variables, not subject capacity. The fact is that the region around 500 nm appears substantially desaturated to these monkeys (see the following), and so when they are initially faced with this test they do poorly. However, with continued training their performance improves and they are eventually quite successful at this task. At any rate, the results from the De Valois and Morgan study, like those of Gunter *et al.* (1965), indicates that *Cebus* monkeys are not dichromatic.

Other tests of *Cebus* color vision carried out by De Valois and Morgan suggest that these animals have color vision similar to that of human protanomalous trichromats. Specifically, measurements of both wavelength discrimination and colorimetric purity discrimination yielded functions that were qualitatively similar to those typically obtained from human protanomalous trichromats.

The measurements of colorimetric purity discrimination for the *Cebus* monkey are shown in Figure 5.6. The question posed in this experiment was how much achromatic light had to be added to each of a number of monochromatic lights in order to render the mixture indiscriminable from an equiluminant achromatic light. Whereas the macaque monkeys produce functions characteristic of normal trichromacy with a single minimum at 560 to 570 nm, the *Cebus* monkeys show a minimum in the region of 500 to 520 nm and, perhaps, a second minimum at about 590 nm. Note also that all the way across the spectrum less achromatic light had to be added to the mixture for the *Cebus* monkeys than for the macaques. The results from this experiment, in conjunction with those described

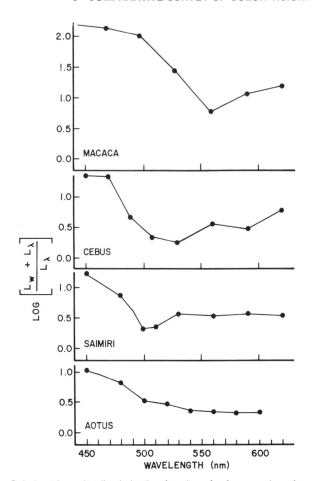

Figure 5.6 Colorimetric purity discrimination functions for four species of monkeys. The plotted points represent the amount of white light that had to be added to each spectral light in order to reduce discrimination performance to 60% correct. (Results for *Macaca* from De Valois *et al.*, 1974; for *Cebus* from De Valois & Morgan, unpublished; for *Saimiri* from De Valois & Morgan, 1974; for *Aotus* from Jacobs, 1977a.)

previously, all argue for the conclusion that the *Cebus* is a protanomalous trichromat.

The squirrel monkey (*Saimiri sciureus*), another very common cebid species, has also been the subject of several color vision experiments. For instance, Miles (1958a) tested these monkeys to see if they could discriminate between several pairs of wavelengths. Although he lacked the information needed to equate these stimuli in luminance, and consequently had to vary their relative luminances over a wide range, it was clear that the presence of color vision could be easily demonstrated in this species. Beyond that, the animals had much greater difficulty discriminating between pairs of long-wavelength stimuli than they did between stimulus pairs taken from other portions of the spectrum. This fact, very probably viewed in conjunction with the Grether results on *Cebus* monkeys, led Miles to conclude that squirrel monkeys are very likely protanopes. As in the case of the Cebus monkeys, this conclusion has also subsequently been shown to be incorrect.

Evidence that squirrel monkey color vision was probably not dichromatic came from several sources. First, five of these animals were tested in a four-choice discrimination task (Jacobs, 1963). As measured in a problem involving the discrimination of flickering lights, the spectral sensitivity of these monkeys was found to be significantly lower in the long wavelengths than it was for subjects having normal color vision. This same outcome also occurs when spectral sensitivity is measured in the context of an increment-threshold task (Jacobs, 1972). However, in color vision experiments, no evidence for the presence of a spectral neutral point was found in any of the five subjects tested (Jacobs, 1963). In this same investigation, color mixing was examined by devising a Rayleigh match test for use with monkeys. The squirrel monkeys were found to require substantially higher proportions of red light in a green-plus-red mixture to match the appearance of the standard yellow light than did normal trichromats. This was taken to imply that the color vision of the squirrel monkey is a trichromatic protanomaly.

Later experiments involving the measurement of colorimetric purity discrimination and wavelength discrimination in squirrel monkeys appeared to support this conclusion (De Valois & Morgan, 1974). In confirmation of the results of Miles (1958a), squirrel monkey wavelength discrimination was found to be most acute in the vicinity of 500 nm. Discrimination was found to be much poorer in the long wavelengths. The results of colorimetric-purity discrimination measurements are shown in Figure 5.6. As can be seen, for the squirrel monkey the spectrum appears maximally saturated in the short wavelengths. Saturation reaches a minimal value at about 500 nm, whereas from 530 to 620 nm, spectral saturation is rather low and roughly constant. The outcomes of both wavelength and colorimetric purity discrimination experiments were consistent with the earlier measurements of the Rayleigh match. All of these results indicated that the squirrel monkey has trichromatic color vision that is defective rather strongly in the protan direction.

In the past few years new evidence has been obtained that shows that the issue of color vision in squirrel monkeys is considerably more complicated than the previous paragraphs indicate. Specifically, it appears that there may be a significant variation in color vision among animals within this species.[2] A strong suggestion that this might be so first appeared in a survey of visual sensitivity among a sample of squirrel monkeys. In that study, increment thresholds were measured at two wavelengths: 540 and 640 nm. Although there was no significant variation in sensitivity to the 540-nm light among 19 animals tested, there was a very large variation (1 log unit) in sensitivity to the 640-nm light (Jacobs, 1977b). Because variations in spectral sensitivity are closely correlated with variations in type of color vision, this result raised the possibility that this species might not be homogeneous with respect to color vision.

Direct tests of color vision in squirrel monkeys have recently been made in my laboratory. Although the variation in color vision among squirrel monkeys can be documented in several ways, the results shown in Figure 5.7 provide a particularly compelling illustration of this variation. Animals are tested in an anomaloscope task, the intent of which is to determine which relative mixture of red and green lights is indiscriminable from a monochromatic yellow light. The top part of Figure 5.7 shows the discrimination performance of two monkeys (solid and dashed lines) for various mixtures of red and green light. The settings on the left-hand side of the scale are for pure red light, those on the right-hand side for pure green, with intermediate settings representing a continuously variable mixture of these two components. Note that both animals successfully discriminated both pure red and pure green from the yellow light indicating that they are not dichromats. Both were also unable to discriminate some mixture of red and green from yellow. That particular mixture represents a Rayleigh match, and its mean location is indicated by the small solid arrows. It is clear that these two animals required significantly different amounts of red and green light to match the standard yellow.

The bottom panel of Figure 5.7 places the results from the squirrel monkey in a comparative perspective. Shown are matches made by a number of trichromatic humans. The range of matches made by normal trichromats is indicated by the extent of the horizontal line. The match points for individuals classified as deuteranomalous are given by D's, those from protanomalous humans by P's. One of the squirrel monkeys required considerably more red light in the mixture than did the normal trichromats. Its match is not different from that of human

---

[2]It has been known for some time that the species *Saimiri sciureus* includes subpopulations that differ phenotypically, and that this phenotypic variation correlates with the geographic origin of the monkey. In recent years it has been established that these different groups have distinctly different karyotypes (for example, Ariga, Dukelow, Emley, & Hutchinson, 1978). It is possible that there may be other, as yet unknown, differences between these various groups of squirrel monkeys, perhaps even some characteristics of their visual systems. In order to reduce potential confusion, it will be necessary for future investigators to specify which of these groups they have studied. In the investigations I performed, all of the squirrel monkeys were of Peruvian origin (''Roman Arch'' phenotype). The same general problem may exist among other species of South American monkeys.

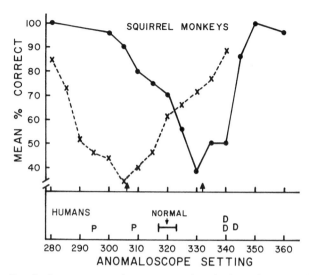

Figure 5.7 Results from an anomaloscope test of squirrel monkeys and humans. The subject's task was to discriminate a variable mixture of red and green light from a monochromatic yellow light. The anomaloscope setting indicates the relative proportions of red and green light in the mixture; pure green lights are present at high settings, pure red lights at low settings. The top panel shows the discrimination performances for two squirrel monkeys for varying proportions of red and green light. Note that they show discrimination failures for significantly different proportions of red and green light (the anomaloscope match values for each are indicated by the small solid arrows). In the bottom panel the locations of anomaloscope match points are shown for human subjects with various types of color vision. The horizontal line shows the range of match values for a number of subjects with normal trichromatic color vision. The match values for several anomalous trichromats are given by P's (protanomalous) or D's (deuteranomalous).

protanomalous observers. This is, of course, quite what would be expected on the basis of all of the earlier work on squirrel monkeys. However, the other squirrel monkey did not require excessive red; instead it selected slightly more green in the mixture than the normal trichromats, although not as much as the deuteranomalous humans required.

The degree of variation in color vision among squirrel monkeys and what significance this variation might hold for normal visual behavior remain to be fully understood. At the time this is being written, what is very clear is that although some squirrel monkeys have protanomalous color vision, not all do. Others that do not have protanomalous color vision also differ somewhat from normal human trichromats, their color vision being slightly displaced in the deutan direction. Still other animals appear truly dichromatic.

A striking confirmation of this within-species variation has been obtained in recent MSP measurements of the photopigments found in squirrel monkey retinas

(Jacobs, Bowmaker, & Mollon, 1981). In those experiments, measurements were made of the absorbance spectra for individual photoreceptors in the retinas of two monkeys. These animals had previously been tested behaviorally; in the anomaloscope test they performed in a manner very similar to the monkeys whose results are illustrated in Figure 5.7, that is, the matches of one classed it as having a severe protan deficiency while the matches of the other were in the deutan direction. Figure 5.8 summarizes the MSP data obtained from these two monkeys. Both had numerous photoreceptors containing a typical rod photopigment and both yielded a few cones having a short-wavelength photopigment ($\lambda_{max} \approx 430$ nm). The only other cone pigment type identified in the severe protan had a mean λ_{max} of 535 nm, a peak location indiscriminable from that for the middle-wavelength photopigment found in the normal trichromatic macaque monkey (Bowmaker *et al.*, 1978). On the other hand, the monkey classed as a deutan appeared to have two classes of long wavelength photopigments with mean λ_{max} of 552 and 568 nm. These results are in accord with the photopigment complements generally believed to underlie these types of defective color vision (Boynton, 1979). They make the reality of the within-species variations in color vision even more compelling, and they provide a presumptive biological basis for that variation.

A particularly interesting New World monkey from the standpoint of its vision is the owl monkey (*Aotus trivirgatus*). This monkey is nocturnal in habit; it is sometimes referred to as the "night monkey." Until fairly recently it had usually been concluded, mostly on the basis of its nocturnality, that the *Aotus* monkey has an all-rod retina and consequently no color vision. However, over the past few years it has become clear that the retina of *Aotus* in fact contains both rods and cones. This retina is still quite atypical of the usual primate construction in two regards. First, the retina contains no clear fovea although there is an *area centralis*. Second, the number of cones found in this retina is much lower than that found in other primates—in the *area centralis* of *Aotus,* cone density reaches about 7300/mm² in comparison to a density of about 48,000/mm² in a comparable portion of the rhesus monkey retina (Ogden, 1975).

Since the retina of *Aotus* appears duplex, and because this animal is often considered to be the most primitive of the New World monkeys in the cebid group, it is of obvious interest to inquire as to the possibility of color vision in this animal. In a brief study, mostly directed toward other issues, Ehrlich and Calvin (1967) were unable to train any of four *Aotus* monkeys to successfully discriminate between spectrally unspecified red and green lights. More recently, both spectral sensitivity and color vision have been examined in a pair of *Aotus* monkeys (Jacobs, 1977a). The photopic spectral sensitivity of these animals was found to be low in the long wavelengths, even lower than it was for comparably tested squirrel monkeys. In a further test involving the discrimination of various monochromatic lights from equiluminant achromatic lights, it was discovered

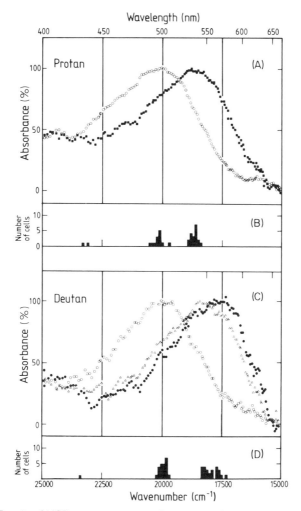

Figure 5.8 Results of MSP measurements made on two squirrel monkeys previously identified as having protan and deutan color vision, respectively. (A and C) Average absorbance
spectra for rods ($\lambda_{\mathrm{max}} \approx 500$ nm) based on 13 receptors for the protan and 24 receptors for
the deutan. In addition, the protan monkey had a single middle-wavelength cone pigment,
that curve being based on an average of 19 receptors; the deutan monkey had two classes of
middle-wavelength photopigments located on the average at 552 nm ($n = 16$) and 567 nm ($n
= 9$). (B and D) Histograms showing the distributions of peak absorbance for all the photo-
receptors. The short-wavelength photoreceptors are represented in the histograms but not in
the plots of absorbance spectra. (After Jacobs *et al.*, 1981.)

that both owl monkeys were able to successfully discriminate between all chromatic–achromatic pairs. This experiment indicated that the *Aotus* monkey has color vision and that this color vision is trichromatic in character.

Further tests showed that although the *Aotus* monkey has color vision, that capacity is not very acute. As evidence for this conclusion, consider the colorimetric purity discrimination data obtained from *Aotus* that is shown in Figure 5.6. It is apparent from the figure that for the *Aotus* monkey the spectrum appears maximally saturated in the short wavelengths, although even there the extent to which lights can be desaturated and still be discriminated from white is substantially less than for normal trichromats. Beyond 500 nm, spectral saturation assumes a low and relatively constant value for the *Aotus* monkey. Indeed, saturation throughout this portion of the spectrum is strikingly low in that only a small amount of white light had to be added to the pure monochromatic light in order to render it indiscriminable from a comparison white light. Still other tests verified the conclusion that *Aotus* color vision, by human standards, is extremely poor. For example, it proved to be very difficult to train these animals in a wavelength discrimination task; the separations in wavelength required for successful discrimination were consistently large, greater than 15–20 nm even at the most sensitive spectral locations. Taking all of these results together, in terms of a formal classification, *Aotus* color vision appears to most closely resemble a very severe protanomalous trichromacy.

To summarize, the color vision of three different cebid species (*Cebus,* squirrel, and *Aotus* monkeys) has been examined in fairly comparable test situations. It appears that although some animals from all of these species have trichromatic color vision, none have normal (human) trichromacy. The nature of the color vision defect in these monkeys initially appeared the same for all three species in that all seemed to be protanomalous trichromats. Beyond that, however, a substantial variation in the degree of severity of this defect was noted. Figure 5.6 illustrates this grading in a test involving discrimination of colorimetric purity. The *Cebus* monkeys show a less severe defect than the squirrel monkeys that in turn have a less severe defect than the *Aotus* monkeys. It must be emphasized again that the gap between squirrel and *Aotus* monkeys appears to be a very large one. In terms of the conventional explanations used to account for the protan deficiencies, it would seem that the long-wavelength photopigment in the cebid monkeys is shifted toward the short wavelengths relative to its spectral position in the normal trichromats.

Recent results require that an important qualification be added to this conclusion. For one of these species, the squirrel monkey, it is now known that there is a significant variation in the type of color vision found among different members of the species. Although this variation is not yet well understood, the possibility that other cebid species might be subject to a similar variation cannot be ignored.

Some data obtained from spider monkeys, to be considered next, makes this possibility even more likely.

Only one other cebid monkey has received any attention at all—the spider monkey (*Ateles*). In his comparative studies, Grether (1939) tested a single animal and found it to have very acute wavelength discrimination, as good as that of the normal trichromat (see Table 5.1). More recently, Glickman, Clayton, Schiff, Guritz, and Messe (1965) also found some evidence for color vision in the spider monkey.

In some recent experiments, color vision was examined in a pair of young spider monkeys (Blakeslee & Jacobs, 1981). The tests were made in the discrimination apparatus illustrated in Figure 2.8. One animal has been found to have excellent color vision, hardly discriminable from that of normal human trichromacy. This monkey required only slightly more green light in a Rayleigh match than did normal human trichromats. In addition, in a test of wavelength discrimination the function obtained from this animal had essentially the same features as those shown by normal trichromats (see Figure 5.9). Both of these tests indicate that this animal has very close to normal trichromatic color vision. However, the results from the second spider monkey tell a quite different story. This animal required significantly more red light in the red + green = yellow test than did either the first spider monkey or normal human trichromats. Furthermore, in the wavelength discrimination test this animal was poorer than the first animal in the long wavelength portion of the spectrum (see Figure 5.9). Its color vision is best described as protanomalous. In sum, this experiment suggests that the within-species variations found for squirrel monkeys may also be characteristic of spider monkeys.

The other species of New World monkey whose color vision has been tested is the marmoset (*Callithrix*). Miles (1958b) presented some evidence suggesting that marmosets do have color vision based on their ability to discriminate between blue, green, and red lights. De Valois and Morgan (personal communication, 1976) have also tested marmosets in the same situation as that in which they examined *Cebus* and squirrel monkeys. Like *Cebus* and squirrel monkeys, the marmosets showed no evidence for a spectral neutral point, and so it appears that the marmoset is also a trichromat. However, again like *Cebus* and squirrel monkeys, it was found that these animals perform in wavelength and colorimetric purity discrimination tasks in a manner suggesting that they have a protanomalous trichromacy. Further documentation of this conclusion is not yet available.

D. Prosimians

In a variety of studies, most of them not very thorough, several prosimian species have been tested for color vision. For example, the galago is a nocturnal

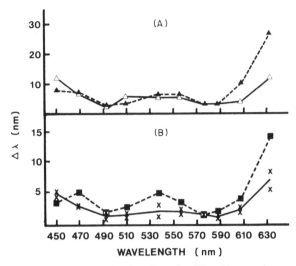

Figure 5.9   Wavelength discrimination functions for (A) spider monkeys and (B) human subjects as determined in the same test situation. The wavelength discrimination values (Δλ) indicate the magnitude of the wavelength change required at each spectral location to support discrimination at a level of 70% correct (chance level = 33%). These values are for changes in both spectral directions except at 450 nm where the change could only be measured toward the middle of the spectrum. For the spider monkeys, subject A1 (△—△) had previously been shown to be a near normal trichromat as judged by her performance on an anomaloscope test; subject A2 (▲- - -▲) was protanomalous. For humans, normals are represented by ×—×, protans by ■- - -■.

animal claimed to have an all-rod retina (Dartnall, Arden, Ikeda, Luck, Rosenberg, Pedler, & Tansley, 1965). Ehrlich and Calvin (1967), in a study referred to earlier, also tested two greater galagos (*Galago crassicaudatus*) to see if they could discriminate red from green lights. Both animals failed the discrimination, but a lack of any clear control of brightness, and perhaps a lack of persistence on the part of the experimenters, makes this experiment hard to interpret. The fact that other nocturnal mammals have recently been shown to have some cones, in addition to large numbers of rods, suggests that a further evaluation of the possibility of color vision in the galago might prove fruitful.

Unlike the galago, a few of the lemur species are diurnal in habit and their retinas contain a mixture of rods and cones. There are several studies on lemur color vision, but none have tried to go beyond the simple question of the mere possession of color vision. Even at that there are some contradictions in outcomes (see De Valois & Jacobs, 1971, for references to some early studies).

In a recent study, Mervis (1974) sought to show that two ring-tailed lemurs (*Lemur catta*) could discriminate between blue, green, and red stimuli. In a two-choice discrimination apparatus these three stimuli were counterposed in

pairs. Starting from a point of human equality the relative luminances of each
pair were varied over a range of about 0.5 to 0.6 log units. Both animals were
able to discriminate between all three pairs, even though substantial training was
required to achieve this success. Although it thus appears that the lemurs have
some color vision, both the extent and nature of this capacity still need to be
determined.

Despite the fact that their taxonomic status is somewhat equivocal, the tree
shrews clearly have sufficient primate characteristics to warrant inclusion of
some comment about them at this point. The common tree shrew (*Tupaia glis*) is
a diurnal, arboreal animal having a retina that is very heavily cone-dominated.
The color vision of this animal has been tested on several occasions. Most of
these studies (see De Valois & Jacobs, 1971, and Mervis, 1974, for references to
these studies) concluded that the tree shrew has some color vision. In the most
extensive investigation to date, Polson (1968) tested five tree shrews on a wide
variety of tests of visual sensitivity and color vision. In measurements of visual
sensitivity she found that the form of the photopic spectral sensitivity function for
the tree shrew is indistinguishable from that of the macaque monkey. On the
other hand, in tests of color vision, Polson found that the tree shrews could
successfully discriminate all spectral stimuli from equally luminant white lights
except for those wavelengths in the immediate vicinity of 505 nm. These results
are illustrated in Figure 5.10. The presence of such a neutral point indicates that
the tree shrew is a dichromat.

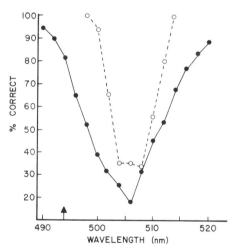

Figure 5.10 Performance in a neutral-point test. The plotted points represent the mean
levels of correct discrimination for five tree shrews (●—●) and one human deuteranope
(○- - -○). The small arrow indicates the wavelength at which an equivalently tested human
protanope showed poorest discrimination. (Data from Polson, 1968.)

Beyond establishing that the tree shrew has dichromatic color vision, Polson conducted several other color vision tests, including both wavelength and colorimetric purity discriminations as well as a determination of various dichromatic color mixtures. The sum of these results, in conjunction with the spectral location of the neutral point (Figure 5.10), and the evidence on spectral sensitivity, led Polson to conclude that the dichromacy of the tree shrew is most similar to that of the human deuteranope. This makes the color vision of the tree shrew unique among those mammals whose color vision has been well characterized.

IV. Comments on the Comparative Survey

Even though this chapter and the previous one contain a restricted review, it should be apparent that the study of comparative color vision has attracted lively attention. It is equally obvious that the picture is far from complete. Not only are large numbers of species entirely uninvestigated, but the experiments that have been done are sometimes flawed or, perhaps more often, sufficiently disparate in technique to make any strong comparisons among them difficult. In the face of these difficulties, a summary classification of the color vision in the species that have been discussed is nevertheless advanced in Table 5.2.

In the course of discussing comparative color vision it has been routine for writers to point out that color vision is a capacity that appears in only selected groups among the vertebrates. In particular, it is often noted that color vision is not widespread among the mammals, and that in this group it only becomes really highly developed among the primates. However, our review makes it clear that, in fact, the *possession* of color vision appears to be very widespread among the mammals. It does not look as if any mammalian species, at least any which have been adequately examined, show a complete lack of a color vision capacity. Having said that, however, it is equally important to emphasize that the degree to which various animals have a color capacity is enormously variant. As we have seen, for many species the color vision that can be demonstrated is quite limited, and perhaps by implication (although surely not proven) not likely to be a very dominant aspect of the sensory world of the animal.

The fact that color vision does vary widely in both quality and quantity across species naturally leads one to wonder if these variations are explainable in terms of the evolutionary status of the animal, or whether the particular type of color vision observed is understandable in terms of the visual demands placed on the animal by its photic environment. These issues are addressed in the next chapter.

Two cautionary points about studies of comparative color vision need to be briefly mentioned. The first has to do with the issue of homogeneity of color vision in nonhuman species. Of course, it is well-known that for both genetic and pathologic causes a significant fraction of the human population has color vision

Table 5.2

Summary Classification of Color Vision[a]

Species norm[b]	Subjects	Comment
Normal trichromacy	Goldfish	
	Great apes	Convincing data only for chimpanzee.
	Old World monkeys	Several macaque species; baboon; African green monkey.
Anomalous trichromacy or dichromacy	New World monkeys	Considerable variation in the severity of the defect among these species. Generalization has been that they are protan. However, there is now evidence for within-species variations with (perhaps) both protan and deutan individuals.
Dichromacy	Tree squirrels	Sciurids are most like human protanopes; tree shrews like human deuteranopes.
	Ground squirrels	
	Tree shrew	
Some color vision (but classification uncertain)	Ranid frogs	Some species possibly trichromatic.
	Turtles	Possible trichromats.
	Pigeon	Excellent color vision, at least trichromatic.
	Cat	Possibly dichromatic. Weak color vision.
	Rat (and other nocturnal rodents?)	Minimal color vision
	Prosimians (Lemur)	Extent of color vision not yet established.
Monochromacy	None	

[a] The species considered are those discussed in Chapters 4 and 5.
[b] The norms are those used to categorize human color vision.

sufficiently different from that of the majority as to be considered "defective." This being the case, one naturally wonders if a similar situation exists in other species. Unfortunately, so far it is hard to know. To establish color vision in other species typically only a small number of subjects are tested, often from two to no more than five. From this small number of subjects a species-general conclusion is usually offered. Given this limitation in sample size, it would be relatively easy to miss a significant within-species variation or, perhaps less easy, to test only minority representatives and thus come to a quite misleading conclusion about the whole species.

In our review of color vision in the platyrrhine monkeys, evidence was presented that shows that there is a significant within-species variation in color vision among squirrel monkeys. And although the data are somewhat more restricted, there is good reason to believe that variations in color vision can also be demonstrated among spider monkeys. In addition, there are some hints of color vision variations among *Cebus* monkeys. For instance, in their studies of *Cebus* monkeys, Gunter *et al.* (1965) reported that individual animals differed considerably in the ease with which they were able to master discriminations between various color pairs. When considered in the context of human color vision, these investigators noted that two of the subjects behaved most like deuteranomalous humans, one behaved more like a protanomalous human, and the other two subjects made errors that did not permit a distinction between these two categories.

In a later study involving measurements of spectral sensitivity, Lepore, Lassonde, Ptito, and Cardu (1975) obtained from a single *Cebus griseus* monkey a spectral sensitivity function indistinguishable from that of a similarly tested normal trichromat. In the context of the previous work on *Cebus* monkeys this led them to suggest that this animal might have a deutan rather than a protan defect. Of course, this result could have equally well led them to suggest that this monkey had normal trichromatic color vision. In either case, the outcome is counter to the idea that all *Cebus* monkeys have protan color vision.

In summarizing these observations on intraspecies variations it can be concluded that, although the data so far are somewhat fragmentary and not entirely consistent, there is nevertheless considerable reason to believe that for at least three genera of New World monkeys there are representative subpopulations whose color vision differs in nontrivial ways. This outcome is at least compelling enough to stand as a clear warning to students of comparative color vision that presuming all members of a nonhuman species to have exactly the same color vision is as unrealistic as we know that presumption would be for all humans.

A final cautionary point concerns the widespread use of the human classification scheme to describe color vision among nonhumans. The difficulty with this tactic is that it presupposes that all color vision can be categorized into those groupings usually found appropriate for human color vision. It is important to remember that each of the categories of human color vision is associated with a number of defining characteristics (for example, color mixing, discrimination data, and color naming), and the implication is that any subject placed into a particular category shows all of these characteristics. This is surely not always the case. To take just one example, the data on the color vision of ground squirrels (reviewed earlier in this chapter) can be interpreted to suggest that these animals have a protanopic dichromacy. However, at the same time, the spectral locations of the neutral points found in these animals (see Figure 5.1) are located at distinctly longer wavelengths than those typically found for human protanopes (indeed, at a spectral location close to that characteristic of human deuteranopes).

Thus, to summarily describe these squirrels as protanopes misleads with regard to some of the characteristics of their color vision. The point is that the use of the human classification scheme should be seen as providing a generally convenient shorthand, and not as a necessarily accurate description of all nonhuman color vision.

V. Results from Studies of Other Mammals

The species considered so far in this chapter obviously represent only a small fraction of all mammalian species. The justification for their inclusion was given earlier. However, questions sometimes arise as to the presence and character of color vision in other species. In this final section some references are provided that will lead the interested reader to what information exists on a number of other mammals.

The results from studies of color vision in assorted mammals are summarized in Table 5.3. As a quick perusal will indicate, the table contains a mixed bag of representatives, some 20 different species including rodents, canids, a number of ungulates, a single marsupial species (the oppossum), and a marine mammal. I expressly make no claim that this is an exhaustive survey, particularly as studies of this kind have often appeared in highly atypical places—as a randomly drawn example, the *Journal of Agricultural Science*. It should also be pointed out that there is sometimes more than one published study for these animals. In those cases I tried to include in the table only the most recent study with the thought that references to earlier studies can be traced through them. Walls' (1942) book should also be consulted for comments on many of the earlier studies.

The dearth of information about color vision in most mammals is probably partly attributable to the nocturnality of many of these species, and to the consequent assumption that most are therefore unlikely to have color vision. Another likely reason is that most mammals are unattractive experimental subjects by virtue of their size or unavailability, or, in some cases, unpleasant disposition. It requires an unusually intrepid scientist to conduct tests of color vision on a subject who in a fit of pique might decide to bodily attack the investigator!

Table 5.3 records the conclusions reached by the experimenters as to whether or not the study yielded evidence for the presence of color vision in the species. These studies vary enormously in quality. Without subjecting all of them to a critical analysis, it is clear that some of them reached conclusions that are, at best, questionable. For example, Kolosvary (1934) concluded that the mouse has color vision based on the fact that this rodent showed a preference for blue strips of paper over red strips of paper for use in nest construction. At this stage it hardly needs pointing out that this behavior could be based entirely on brightness differences. In another case it was concluded that a fox had no color vision

Table 5.3

Results of Studies of Color Vision in Assorted Mammals

Animal[a]	Presence of color vision[b]	Reference
Mouse		
Mus musculus	Yes	Kolosvary (1934)
Apodemus sylvaticus	No	Salzle (1936)
Clethrionomys glareolus	Yes	Salzle (1936)
Guinea Pig: *Cavi porcellus*	No	Miles *et al.* (1956)
Rabbit: *Oryctolagus cuniculus*	Yes	Nuboer (1971)
Raccoon: *Procyon lotor*	No	Michael, Fisher, & Johnson (1960)
Opossum: *Dildelphis virginiana*	Yes	Friedman (1967)
Fox: *Vulpes vulpes*	No	Osterholm (1964)
Dog	Yes	Rosengren (1969)
Swine	Yes	Klopfer & Butler (1964)
Sheep	No	Tribe & Gordon (1950)
Nilgai antelope: *Boselaphus tragocamelus*	Yes	Backhaus (1959a)
Pigmy goat: *Capra hircus*	Yes	Backhaus (1959a)
Domestic goat: *Capra hircus*	Yes	Buchenauer & Fritsch (1980)
Red deer cow: *Cervus elaphus hippelaphus*	Yes	Backhaus (1959a)
Giraffe: *Giraffa camelopardalis tippelkirchi*	Yes	Backhaus (1959b)
Zebu	Yes	Hoffman (1952)
Horse	Yes	Grzimek (1952)
Cow	Yes	Dabrowska, Harmata, Lenkiewicz, Schiffer, & Wojtusiak (1981)
Seal: *Phoca largha*	Yes	Wartzok & McCormick (1978)

[a] Specific designations are supplied in those cases where the original author did so.
[b] The conclusion (yes or no) is the one reached by the author of the study as to whether or not the animal tested was shown to have color vision.

because it was unable to learn an association of red = food and blue = no food (Osterholm, 1964). In this experiment the animal was trained for "two evenings" prior to testing. As I have noted in other contexts, demonstration of color vision in some species may require considerably greater persistence on the part of the investigator than this study suggests.

Almost without exception, the studies listed in Table 5.3 were only intended to determine whether or not the subject possessed color vision. Consequently, no conclusions can be derived as to the type of color vision these various species possess. It is also hard to gain much feeling for the acuteness of color vision in these animals, partly because determinations of wavelength differences needed for discrimination were seldom measured, and because most of these experiments used stimuli whose spectral specifications were not provided. Despite this, it is clear that few of these subjects have "good" color vision, at least as judged by the usual human norms. One possible intriguing exception to this generalization is the swine. In a study reported only in abstract form, it is claimed that swine were able to master a considerable number of color discriminations, and were successful at discriminating wavelength differences of less than 20 nm around 575 nm (Klopfer & Butler, 1964).

Because of their special relationship to man, some comment should be made on the question of dog color vision. The study by Rosengren (1969) represents the most modern report discovered. In that paper, sixteen earlier studies of dog color vision are listed. Of these, Rosengren reports that nine studies concluded that dogs do have color vision. Rosengren's own experiment was conducted on cocker spaniels, in the author's home, and with the simplest of equipment and procedures. She taught her dogs to select one from among six dishes. These dishes were painted in shades of gray or blue, green, yellow, red, and orange. Each of the colored dishes was one of a series in which the hue varied from light to dark. First she found that the dogs were capable of errorlessly picking a red dish from among a series of gray dishes. Then, in later experiments, it was found that the animals could discriminate between dishes of various hue combinations, and between each of the colored dishes and a series of gray dishes. A number of procedures were used to eliminate extraneous cues. The experiment, although somewhat casual by the standards of many of the other experiments we have discussed, appears convincing in its conclusion that dogs can make a variety of discriminations based on color. Perhaps the recent interest in developing a number of canine models for the study of retinal disease states will encourage someone to carry out a comprehensive investigation of dog color vision.

Chapter 6

Evolution of Color Vision

The two previous chapters should have made it abundantly clear that color vision varies widely in both quantity and quality among the vertebrates. Sooner or later that fact alone compels one to think about the adaptive advantages of this sensory capacity. A number of questions might be raised. For example, why have some species evolved good color vision while others have not? What are the particular environmental circumstances that reinforce the retention of those alterations in nervous systems required to extract color information? Is there a definable sequence in which these nervous system changes have occurred? Can one identify contemporary examples of color vision systems that may be illustrative of different evolutionary stages? None of these questions can presently be answered with certainty. Nevertheless, if not answered, some comments, often highly speculative and certainly arguable, can be directed toward them. That is the goal of this final chapter.

I. The Utility of Color Vision

What is the purpose of color vision? The presence of diverse pigmentation in our natural world creates an environment in which the spectral energy distribution reaching the eye of a viewer varies from location to location in space. As we have seen, a capacity to extract information about the differences in these distributions, irrespective of their absolute energies, constitutes color vision. The perceptual result, for man and other species having color vision, is a multihued world in which objects appear to merge and contrast by virtue of their differences in color. It is hardly surprising, therefore, that color vision has typically been viewed as a means toward more efficient detection of particular visual stimuli, or of better discrimination among them.

In considering why species other than man possess color vision, Walls (1942) suggested that because these species "cannot appreciate sunsets and old masters" (an assertion that the recent work on chimpanzees and Old World monkeys reviewed in Chapter 5 now makes somewhat less certain), then there must be some other reason why they have evolved color vision. That reason, Walls went on to conclude, was that color vision "promotes the perception of contrast and hence, visibility.'' This view, that the purpose of color vision is to enhance visibility and thus the detection of an object from its surroundings is directly endorsed or implicitly assumed in many discussions of animal behavior. A common illustration of the kind intended to convince us why this must be so is given by Polyak (1957) who published in his monumental book on the vertebrate visual system a vivid colored picture of a bush containing both ripened and unripened fruits, the ripe red fruits appearing highly visible amonst the green leaves while the unripe green fruits are considerably less obvious. Polyak (1957) goes on to suggest that not only does this color difference permit birds to discriminate ripened from unripened fruits, but may also be mutually beneficial to both species in that birds are attracted to a food source and then serve as agents to disperse plant seeds to other locations, thus providing the plants with an adaptive advantage as well. Indeed, this tie is argued to be so important that it is speculated that color vision and colored fruits may have co-evolved (Polyak, 1957).

Although this view may appear intuitively obvious, it is nevertheless instructive to consider the issue a little further. Exactly what evidence is there that the presence of color vision is to "promote the perception of contrast"? To make this question a little more tractable let us arbitrarily suppose that there are three sequential stages involved in object perception in which the role of color vision might be examined: object detection, object recognition, and the signal properties of color. These correspond, roughly, to the following perceptual descriptions: red blob on a green background (object detection), red apple among green leaves (object recognition), ripe apple (signal properties of color). In the next three sections some comments will be made on each of these issues.

A. Object Detection

To start thinking about how color vision might aid in object detection, it is useful to consider for purposes of illustration a very simple (and highly artificial) visual scene, a single barshaped figure superimposed on an extended homogeneous background. The bar may differ from its background in luminance, in chromaticity, or in both of these dimensions. If the two differ only in luminance, as, for example, a white bar on a dim gray background, then the difference between the two can be specified as luminance contrast: the ratio of the difference in luminance between the bar and its background relative to the sum of the luminances of the two. Other things equal, in this situation the higher the contrast so computed the more distinct the bar will appear. Obviously, this relation holds whether the viewer has the use of color vision or not.

Alternatively, the bar and the background region could be made equal in luminance but different in chromaticity. One dimension along which the bar and its background may differ is in their colorimetric purities. For example, suppose the bar is made up entirely of wavelengths between 640 and 660 nm whereas the background contains a mixture of all spectral wavelengths—a red bar on a "white" background. In this case, if the red bar and the white background are equal in luminance then the two will be indiscriminable to a viewer without color vision, but will be easily seen by an observer with color vision. If white light is progressively added to the red bar (while at the same time progressively removing some red light from the bar so as to keep it at the same luminance), then the bar will become continuously less distinct until, at some low purity level, the bar will become indiscriminable from the white background. In this case our ability to see the bar depends entirely on the presence of color vision, and how visible the bar is will be a function of the difference in colorimetric purity between the bar and its background, and on its wavelength (since, as noted in Chapter 2, saturation depends on both of these variables).

A third possible arrangement in this situation is where the bar and its background are equal in luminance and in colorimetric purity but different in wavelength content—for example, a green bar on a red background. Again, in this case, the bar will be visible only by virtue of the color vision of the viewer. Here the degree of its visibility will depend on the difference in the wavelengths of bar and background. Obviously, this situation is only a slightly exotic version of the familiar wavelength discrimination task. Thus, if the bar is 530 nm and the background is 535 nm, it will be harder to discriminate between the two than if the background were changed to, say, 545 nm. And, as in the case of alterations in colorimetric purity between the bar and background, the discrimination of pure-wavelength differences will depend on the magnitude of the differences in wavelength between the two and on the particular wavelength pairs involved (as documented in Figure 2.11).

In these simple examples of a single bar on a homogeneous background it can be easily appreciated that the presence of the bar may be detected by virtue of luminance differences between the bar and its background, or as a result of what we may call "pure-color" differences. These latter situations are examples of instances where the presence of color vision not only enhances the possibility of detection, but where in fact it is the only thing that makes detection possible. Of course, the reader may quite reasonably object that these conclusions hardly need documenting. Obviously, our ability to discriminate form in commonplace situations containing only luminance differences, like black and white motion pictures, is frequently superb. And although natural scenes containing only color differences may be considerably harder to find, bright multicolored drawings can approximate that situation, and it is obviously also easy for us to discriminate form in those instances.

Is it possible to say exactly how much, and under what conditions, object visibility is improved by the utilization of color vision? Can one supply measurements to answer these questions? At least theoretically it should be possible to do this, to ask, for example, how much more visible a target becomes if chromatic differences can be used as well as luminance differences, or vice versa. This has not been done except under very restrictive conditions, far from those situations where animals must normally utilize color vision. However, some of the considerations that may be relevant can be identified, and it will be useful to do so here.

First, the relative detectability of one kind of form stimulus (albeit a monumentally uninteresting one from the viewpoint of an observer), a grating pattern, has been systematically investigated under conditions where the grating pattern contains either only luminance differences or only color differences. The basic idea is to determine for any given spacing of these gratings (usually composed of spatially repetitive sine waves that are described in terms of numbers of cycles per degree of visual angle) the minimal contrast needed for the grating to just be discriminated from a similar, homogeneously illuminated field. A typical result of such a measurement is illustrated in Figure 6.1. It can be seen there that the detectability of the grating depends on the spacing of the bars. In this instance the grating can be discriminated at a minimum level of contrast when the pattern has a period of 5 to 10 cycles/degree (equivalent to a bar width of from 3 to 6 minutes of visual angle). Not surprisingly, higher luminance contrast is required to reach the same level of detectability as the grating is made finer and finer until a point is reached (arrow in Figure 6.1) where the grating can no longer be

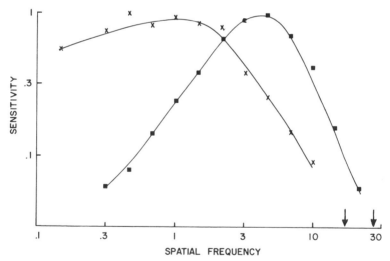

Figure 6.1   The contrast sensitivity of human subjects for sine wave gratings. The spatial changes were defined by either pure hue differences (×) or pure luminance differences (■). The small vertical arrows indicate the approximate maximal spatial frequencies to which these subjects were sensitive for both kinds of stimuli. (Data replotted from De Valois & Jacobs, 1981.)

discriminated even at maximum luminance contrast. This point corresponds to a threshold visual acuity, its exact value depending on a wide range of other situational variables (adaptation state, luminance level, retinal location, etc.). Under photopic light levels, and with foveal viewing, this value might be on the order of 30 to 60 cycles/degree. Note also that the luminance contrast required to detect very coarse gratings is also higher than that for intermediate values—that is, higher than a luminance grating (and presumably other figures defined by luminance differences) whose components are of some intermediate size.

The alternative situation, that where the sinusoidal grating pattern is made up of color variations but is without luminance variations, for example, a grating which in one cycle changes continuously from red to yellow to green without any luminance changes, has not been investigated nearly as intensively. What has been established (van der Horst & Bouman, 1969; Granger & Heurtley, 1973) is that the results of similar determinations for pure-color changes are significantly different from the picture just outlined for luminance gratings. Specifically, the very highest spatial frequencies that can be detected in pure-color gratings are considerably lower than for those gratings that have luminance variation (Figure 6.1). In addition, there is almost no relative loss in sensitivity at the low spatial frequencies for pure-color gratings analogous to those seen for luminance gratings.

What do these experiments on sensitivity to grating patterns indicate about the determinants of the detection of visual objects? First, the differences in sensitivity to gratings made up of either luminance or color variations indicate that fine spatial resolution is better when the differences between spatial locations include luminance changes than when they just include color changes. Indeed, one of the hallmarks of man's vision is that visual inputs must be summed over larger amounts of time and space in order to be effective when these inputs have only color variations than they do when the inputs involve luminance variations. Second, the relative differences in sensitivity to the low spatial frequencies for luminance and color changes may be taken to imply that the contrast between neighboring regions in space is stronger in the color domain than it is in the luminance domain; that is, very gradual spatial changes in color will be more detectable than equally gradual changes in luminance.

The last point, that there may be differences in "contrast" between achromatic and chromatic vision needs further discussion. Thus far in noting how differences in luminance or in the spectral energy distribution might contribute to the detection of an object we have talked as if the issue can be understood merely by considering the differences along these dimensions in various parts of a visual scene, as in the example of the bar and its background. Real visual systems, however, do not separately process inputs from adjacent locations in space. Rather, beginning at the initial stages in the visual system, the analysis of inputs from adjacent locations is a comparative and interactive one. Consequently, the effects of inputs from one location may greatly influence the effectiveness (and hence the resulting perception) of inputs from neighboring locations. In some cases these interactions enhance the perception of differences between two locations, in other cases diminish the perception of differences. In either instance, they need to be taken into account in trying to determine how color vision may contribute to the detection of real objects.

Before talking further about these effects, it may be worthwhile to observe in passing that the term "contrast" is commonly used to refer to two different situations. Thus, any spatial regions that appear different are described as "contrasting." In addition, however, there may be some active nervous system processes that serve to enhance the perceived differences between such regions, that is, to make the regions appear more different when they are spatially juxtaposed than when they are viewed in isolation. This enhancement process is also commonly called "contrast."

To return to the discussion of how color vision may contribute to object detection, let us consider contrast as an enhancement process. I have noted that the appearance of a visual stimulus may be altered by the spatial juxtaposition of another stimulus. That change in perception is often in the direction of making the perceived differences between two objects greater when they are viewed together than when either are viewed separately. Thus, if in the first example, the bar is of a low luminance and it is viewed on a background of slightly higher luminance it appears as a dim gray. However, if the luminance of the background is now raised to a higher level, the bar appears black. This is a familiar example of simultaneous brightness contrast. It is often a very powerful effect that tends to make stimuli of different luminances appear even more different in brightness when the two are spatially juxtaposed. Of course, an analogous effect can also occur where there are temporal variations in luminance, so-called successive contrast. These effects have been much investigated, both with regard to psychophysical relationships and to their physiological bases.

Contrast effects of a similar nature also occur in the color domain. When placed on a red background, the gray bar tends to take on a greenish tinge, or it becomes reddish if the background is made green. In general, this change, simultaneous color contrast, operates in the direction of producing a hue that is roughly complementary to that of the inducing light—reds induce greens, yellows induce blues, etc. The usual textbook demonstrations of simultaneous color contrast suggest that it is a rather negligible effect: the red background makes the gray bar only slightly greenish, etc. Because of this, it might be supposed that these effects are of little practical importance. However, Land (1959) has shown that in complex visual scenes, like most natural environments, color contrast effects can be very substantial. This observation brings up an important point; namely, that contrast effects are very much dependent on the context in which they are assessed. Whereas the appearance of a single bar on a homogeneous background may be relatively easy to predict, it may be quite a different story when stimuli that carry added significance to the observer are involved. As an example of this fact, Dodwell (1975) points out that the amount of green induced into a small achromatic region by a red surround depends on the shape of that small region; if the outline is that of a leaf it will appear considerably greener than if the outline suggests the shape of a donkey!

Paradoxically, spatial juxtaposition of stimuli that vary in luminance and chromatic content may also generate changes in a direction just opposite to those described as contrast effects. For example, under some circumstances the gray bar seen on the red background may take on not the greenish tinge predicted by contrast, but rather a reddish tinge. Effects of this kind, often collectively called *assimilations,* have not been nearly as intensively studied as the contrast effects. However, it can be concluded from studies of these effects that whether contrast or assimilation (or, indeed, no effect at all) results from the spatial juxtaposition of stimuli depends on quite a number of characteristics of those stimuli—relative sizes, luminances, etc. (Helson, 1963)—as well as on more complex variables such as the attentional state of the observer and like issues. Probably the main preliminary conclusion to be drawn here is that it is simply not appropriate to assume that a difference in chromaticity between two regions automatically makes them more discriminable. Indeed, quite the opposite may occur.

B. Object Recognition

Object detection and object recognition are interwoven processes. Obviously, one must be able to detect an object in order to recognize it. Some of the features contributing to detection mentioned in the previous section also bear on recognition, just as some of the aspects here described under "recognition" may contribute to object detection. To that extent, the separation of this discussion may be an artificial one.

Two regions in visual space may appear different, may contrast, by virtue of differences in luminance or chromaticity between the two regions. However, if the spatial zone of transition between these two regions is gradual rather than abrupt, our ability to discriminate between them, and perhaps more importantly to identify the shape of the components, a possible clue to object recognition, may be poorer than if the transition is a crisp one. Because this is so, these regions of transition, commonly called borders or contours, may be of particular importance in the discrimination of form.

The formation of contours by luminance differences has been much investigated. In such situations, brightness discontinuities are frequently seen at the border—that is, on the two sides of the contour, bands appear that are brighter and dimmer than are the regions located further from the contour. These discontinuities are usually referred to as Mach bands. The characteristics of these bands, the stimulus parameters on which they depend, and the physiological models that have been invoked to account for them have been extensively reviewed elsewhere (Ratliff, 1965). Suffice to say here, the occurrence of Mach bands at the locations of luminance discontinuities should serve to make the shapes of objects more distinct by highlighting their outlines.

The question, however, is not how spatial luminance variations create contours but rather how chromatic changes may do so. A first consideration is whether or not the brightness discontinuities that are useful for discriminating contours formed by luminance changes have a chromatic counterpart. The issue of whether or not so-called "colored Mach bands" occur at all is unsettled. Part of the problem in deciding whether chromatic changes can produce Mach bands is entangled in the purely technical problem of generating chromatic borders that have no luminance variations, while part may be due to differences in the definition of just what constitutes a colored Mach band (these issues have been discussed by Pease, 1978). Despite this uncertainty, it is clear that whereas border enhancement effects can be easily produced by luminance changes, it is, at best, very difficult to do the same with chromatic changes alone. Most convincing to me in this regard is the careful investigation by Tansley and Boynton (1978) in which it appears clear that there is a lack of perceived "distinctness" when a chromatic border lacks any luminance discontinuities.

The study just referred to represents part of a linked series of investigations carried out by Boynton and his colleagues (reviewed by Boynton, 1973) that were directed toward providing quantitative measurements of the degree of distinctiveness of a border formed between two adjacent and homogeneous regions that vary in luminance, in chromaticity, or in both. In addition to determining that no border enhancement occurs at the boundary between two regions of different chromatic content but of equal luminance, a fact already noted, it was also demonstrated that the strength (that is, distinctiveness) of the border between two such regions depends, for the human observer, entirely on the relative activities of the cones containing photopigments having λ_{max} at 536 and 565 nm. That is, the activity of the cones containing the short-wavelength photopigment ($\lambda_{max} = 420$ nm) do not contribute to the formation of chromatic borders and, thus, as far as the formation of chromatic borders is concerned, the normal human trichromat behaves much like a tritanopic dichromat (Tansley & Boynton, 1978).

This finding accounts for a number of qualitative observations. For instance, it had been reported by earlier investigators that when certain pairs of equiluminant stimuli of different chromaticities are placed side by side one fails to perceive a distinct border between them. Rather, in these cases the two fields appear to merge into one another with the boundary between them vague and, sometimes, completely indiscriminable. Tansley and Boynton (1978) argue that each of these involves a situation where there is no differential stimulation of the 536- and 565-nm cones, such as the spatial juxtaposition of equiluminant blue and gray regions. In these situations the shapes of the two regions may become entirely indistinct.

For our purposes, the important implication of these observations is that although the 420-nm cones contribute strongly to the perception of color in the sense of signaling hue and saturation, they do not contribute to the formation of chromatic borders. The reasons why this is so are not known for certain, but there is evidence to suggest that it results from the relatively low spatial and temporal resolution observed in the neural signals originating from the 420-nm cones (Tansley & Boynton, 1978). The extent to which there are classes of cones in other species that have the properties of the 420-nm cones in man cannot yet be evaluated (although the short-wavelength cones in other nonhuman primates and in the cat would appear to behave analogously as was noted in Chapter 3). Certainly in those species described previously that have color vision but only relatively small proportions of cones, it would be reasonable to suspect that those cones are unlikely to contribute much toward the formation of clear chromatic borders.

Sharp borders and contours are composed of high spatial frequency components. Thus our inability to perceive sharp borders that are created by pure-color changes reflects directly the lower spatial resolution of the color vision system that is documented in Figure 6.1. Despite the relative inadequacy of the color vision system in resolving sharp boundaries, there is some evidence to indicate that this by itself may not prohibit the color vision system from contributing centrally to the recognition of objects. For example, Ginsburg (1976) has examined the abilities of human subjects to recognize familiar objects that have been frequency filtered so that various frequency components are removed from the stimulus. He finds that object recognition for simple and even complex forms, such as human faces, remains high even when the high-frequency components have been completely removed. In his view, the lower spatial frequencies provide information about objects while the higher frequencies contribute to the perception of fine detail. Thus, although relatively insensitive to the higher spatial frequencies, the color vision system could still contribute significantly to the recognition of objects. Indeed, it has recently been noted that there are instances where sensitivity to high spatial frequencies could prove to be a disadvantage, where such sensitivity might make it more difficult to perceive form (De Valois & De Valois, 1980). In such cases the color vision system might have an advantage over the luminance system by virtue of its lower spatial sensitivity.

It is apparent that there are a large number of facts potentially relevant to the question of how the presence of color vision might influence the detection and recognition of objects in the environment, of which the foregoing represent only a sample. Do these observations lead to any certain conclusions? In the following paragraphs I attempt to summarize these observations and to suggest a few additional considerations central to this issue.

First, it is quite certainly the case that color vision will make possible the detection of objects when these objects differ from their surroundings only in chromaticity and not in luminance. Situations of this kind in natural environments are rare, but to the extent that they occur at all, then a color vision system is a definite asset in the discrimination of form.

Second, we have noted a number of experimental situations where the capacities of the color vision system have been tested in isolation. They show, in sum, that

1. the color system has lower spatial resolution than does the black–white system in an animal that has both,
2. the border enhancement effects so characteristic of luminance borders do not appear to have a color analogue, and
3. not all components of a color vision system are equal in their contribution to the formation of chromatic borders.

To the extent that these properties yield predictions about the discrimination of objects, then color vision systems per se have some limitations as devices for making discriminations in visual space. However, it is important to remember that for gross recognition of objects, the lower spatial resolution of the color system may not be at much of a disadvantage.

Third, there are spatial interactions within color vision systems that may contribute to the perception of the environment. Some of these tend to enhance the effects of spatial differences in chromaticity, but there are also spatial interactions based on color vision (assimilation effects) that tend to make it less likely that neighboring regions will appear different and thus be discriminated. Which, if either, of these two effects occurs appears to depend on many aspects of the viewing situation. In any case, it is simply not justified to automatically assume that spatial interactions in the color domain enhance form discrimination.

In attempting to understand how the addition of color vision to a visual system might make the system better able to discriminate objects in the environment, it is impossible to ignore the fact that vertebrate visual systems, whether they include a strong color-discriminating capacity or not, are remarkably adept at discriminating luminance differences. The degree to which they can do so depends, of course, on a myriad of other variables. Even so, I am not aware of any species that fails to show an ability to discriminate small luminance differences under some sets of stimulus conditions. Indeed, the power of luminance differences as cues for discrimination is so great that in many instances an observer will spontaneously use these luminance differences as the relevant dimension for discrimination, even when others are available. This is particularly manifest in color discrimination tests, as the discussion in Chapter 2 indicated. In testing large numbers of subjects from various mammalian species, some of which have ''good'' color vision, I have often observed that given a luminance difference as a cue, the animal frequently uses that as a basis for discrimination even if a color difference known to be discriminable is also available.

The comments in the preceding paragraph should not be taken to indicate that color vision does not foster the discrimination between objects. Rather, all they seek to show is that vertebrate visual systems appear superbly well designed to discriminate luminance differences, and these luminance differences provide a very powerful cue for the discrimination between stimuli even when other differences are available.

There is a large body of experimental results bearing on the stimulus dimensions used in the discrimination of form by human subjects. Because luminance and chromaticity differences are typically not clearly separated in these experiments, they do not provide much additional leverage on the question we are examining. However, these studies do make some additional points. For one thing, experiments have shown that the time required to detect a particular object in a pictorial display is significantly shortened (by as much as one-third) if the display is rendered in color versus the time required if the same display is viewed in black and white (Christ, 1975). A difference of this magnitude implies a powerful advantage for the system having color vision versus one that does not, and this in itself might constitute a prime reason for its evolution.

Another relevant finding is that multiple cues for discrimination are frequently not simply additive. Thus, the combination of two equally powerful cues for the discrimination of an object does not render the object twice (or any other single factor) as discriminable as it was when only one of the cues was present. Indeed, the presence of the second cue may actually decrease the discriminability of the first (Garner, 1974). In the context of the present discussion this could mean that the addition of a discriminable color difference might decrease the discriminability of a concurrent luminance difference, or vice versa. That this can in fact happen is commonplace: a luminance difference between two lights of the same spectral content is very easy to discriminate, but if one of the lights is now changed in spectral content while the luminance relationship between the two is not changed, then although the two may continue to be discriminable by virtue of their color differences, the luminance difference may become considerably less obvious. The point is that by adding color vision one may change the perceived nature of the differences between objects without necessarily making the two ''more discriminable.''

In seeking to understand how the presence of color vision might ''promote visibility,'' the preceding discussion has been focused on stimulus situations that are artificially contrived, that is, that have few parallels in the real world. In point of fact, of course, objects differ from their surroundings not just in luminance or just in chromaticity, but most typically in both. And this brings us back to the initial question. If objects usually differ from one another in luminance, and if all visual systems have the capacity to effect good luminance discriminations, why has color vision evolved? The answer we have been examining is that color vision promotes the perception of objects, and there is much evidence that this is true. However, it cannot necessarily be assumed that color vision evolved solely as a device to promote object detection and recognition. In addition to those roles, color vision provides another dimension of information about the visual world, and it is that aspect that now requires brief attention.

C. Signal Significance of Color

It is obvious that quite beyond our ability to discriminate between objects through the use of color vision, this sensory capacity also provides an additional dimension of information about the visual environment. That is, color vision yields a set of categories of color experience that may be used as an additional means of communicating information. This additional use of color vision, what might be called its signal significance, could also be viewed as a prime benefit in the evolution of color vision systems. In essence, the distinction between these two uses of color vision is what, for the human at least, corresponds to wavelength (or colorimetric purity) discrimination versus color naming. Although the ability to name colors is clearly predicated on the discrimination capacity, it goes considerably beyond it in terms of the information it yields.

The signal significance of color for humans is important and obvious. Commonly cited examples are the use of color to judge the edibility of fruits, or the physician's use of skin color as a diagnostic to state of well-being. In these cases color provides information about the internal quality of an object whose external form is unchanging. There is little concrete evidence that this second means of utilizing color vision has strong importance in species other than man. However, the color naming and color preference studies carried out on nonhuman primates and birds reviewed in the previous chapters at least permits the inference that color can be employed as a categorizable continuum by other animals that have color vision. There is also a substantial literature bearing on the significance of animal coloration as a means of communicating visual information (summarized by Hailman, 1977). Obviously, animal coloration patterns are utilized as a means of enhancing or diminishing the visibility of the animal (that is, along the lines of using color vision like those described in the previous two sections). However, it is equally likely that in many instances these chromatic patterns are more important in the visual transference of specific information, rather than just as a means of altering visibility.

Speculating on the utility of any evolved characteristics is obviously a very risky business in which any number of alternatives can be made to seem eminently reasonable. It is no less so in the case of color vision, perhaps even more, because in doing so we have been discussing color vision as if it can be considered separately from other aspects of vision. Of course it cannot, and thus the reasons for the evolution of color vision will eventually have to be examined in the context of the entire visual system and its range of capabilities. Despite this caveat, it does seem reasonable to suggest that the utility of color vision may fall under two general headings which have not, I believe, been sufficiently separated in speculations of these kinds.

These two different roles for color vision may already be seen to have some physiological counterparts. In Chapter 3 it was noted that there are some cells in the monkey visual cortex, so-called "multiple color cells," whose responses indicate that they are utilizing color information as a means toward making spatial discriminations but without reporting any specific information about which colors are involved. These cells are the kind one might expect to find exclusively if the only role of color vision was to make spatial discrimination more efficient. However, it was additionally noted in that discussion that cells have also been found that respond quite specifically to different colors. Some of the latter found in Visual Area IV appear to be greatly concerned with the color of the stimulus, but not much concerned with its particular spatial characteristics (Zeki, 1973).

II. Context for the Evolution of Color Vision

The previous section contained some suggestions as to the ways in which color vision may offer sensory advantages to its possessor. The degree of success that an animal has in its interactions with the environment is a key factor in determining whether nervous system changes offer adaptive advantages or not. If they do, such changes are selected and retained. In the case of color vision, therefore, one should expect to find specific environmental features that have guided the evolution of the neural machinery necessary for this capacity. What might these factors be?

It may be useful to approach this issue from the negative side. That is, to ask why all species do not have good color vision. One possible answer to this can be derived from a consideration of the nature of the physiological organization underlying color vision. The argument supporting this suggestion has recently been given elsewhere using the primate visual system as an example (De Valois & Jacobs, 1981).

The gist of this argument is illustrated schematically in Figures 6.2 and 6.3. Figure 6.3 shows two stimuli, one of which is a pure-color grating (Figure 6.3A) composed of red and green sinusoidal stimuli so positioned that there are no luminance variations across the pattern although there are continuous chromatic-

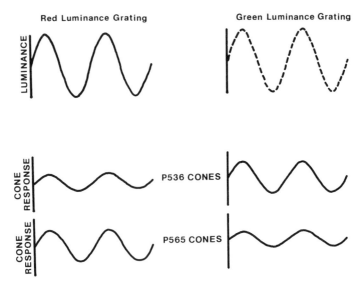

Figure 6.2 Illustration of the relative magnitude of the responses that would be elicited from two cone pigment systems (λ_{max} = 536 and 565 nm) in response to two different spatial stimuli: a red grating varying sinusoidally in luminance and a green grating varying sinusoidally in luminance. Note that both cone types respond to both stimuli, but to slightly different extents. (After De Valois & Jacobs, 1981.)

ity variations over this same region. If these same two stimuli are positioned so that they are in phase (Figure 6.3B), then the resulting stimulus contains luminance variations across the pattern but no variation in chromaticity. Next, recall from Chapter 3 that the two long-wavelength cone pigments in the primate retina (λ_{max} = 536 and 565 nm) have considerable spectral overlap and that they generate electrical signals that are proportional to the number of quanta they absorb. These facts are schematized in Figure 6.2, which shows that these two cone pigments are both activated (but to slightly different extents) by a red grating varying in luminance and a green grating varying in luminance (the separate components of the stimuli shown in Figure 6.3).

Given these cone responses it can be seen that summing the two receptor outputs will provide a large differential signal in the case of the pure-luminance grating, but no differential response in the case of the pure-color grating (compare Figures 6.3C and 6.3D). On the other hand, the difference between the outputs of the two pigment systems will yield little or no differential response in the case of the pure-luminance grating but will give a differential response to the pure-color grating (compare Figures 6.3E and 6.3F). Since Figure 6.3A defines an appropriate test for color vision, this is just another way of demonstrating that the presence of two cone pigments is a necessary but not a sufficient basis for

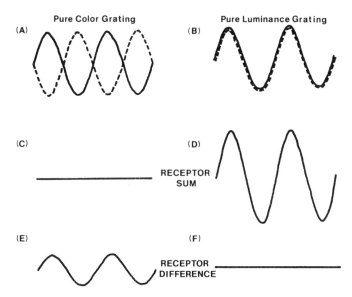

Figure 6.3 (A and B) Sine wave stimuli formed from a combination of two sinusoids illustrated in Figure 6.2 (- - -, green; —, red). In (A) the two have been superimposed out of phase so that the resultant contains chromaticity changes but no luminance changes. In (B) the two have been added in phase so that the sum has luminance variation but no chromaticity variations. Traces (C) through (F) illustrate the kinds of responses that would be obtained from cells connected to the 536- and 565-nm cones for these two stimulus configurations under situations where the receptor outputs are either summed or differenced. See the text for further discussion. (After De Valois & Jacobs, 1981.)

color vision—in addition, one must have some means of comparing the relative effectiveness of different spectral lights on the two pigment systems.

The main point of this example, however, is to show the relatively small size of the difference signal as opposed to the summative signal. This smaller size, of course, is a direct consequence of the fact that the relevant cone pigments have considerable spectral overlap and thus the differences generated by any given stimulus will be quite small relative to the sum of its effects on the two pigments. Consequently, it will require considerably higher light levels to produce a criterion level signal in the differencing system than it will to produce the same sized signal in the summing system. To the extent that the primate system is a good model, and this may be an arguable supposition, this suggests that animals whose photic environments are primarily nocturnal might be expected to have good luminance discrimination (the summing system) but poor color vision (the differencing system). There are admittedly debatable assumptions, but this does provide one suggestive basis for understanding why all species do not have good color vision: color vision systems require a considerable amount of light for efficient operation and this is not present in the normal environments of many animals.

Given that the association between color vision and a diurnal lifestyle may be understandable in the context of the mechanisms vertebrate nervous systems employ to extract color information, this still leaves the larger and more interesting question of why color vision is so variable in its presence and quality among diurnal species. The availability of reasonably high environmental light levels may be a necessary condition for the evolution of color vision, but that is clearly not sufficient to guarantee good color vision. For example, one can hardly imagine a more resolutely diurnal animal than the ground squirrel who, as Walls (1942) succinctly pictured it, "unblinkingly tolerates intensities which force us to screw up our eyelids or run for a pair of dark goggles." Despite this, as we saw in the previous chapter, this animal has relatively poor color vision. Obviously, there must be environmental features beyond the sheer quantity of available light that are critical in the evolution of color vision.

Although it is hard to believe that color vision systems did not evolve in concert with the particular spectral energy distributions that are critical that each species be able to discriminate, not much firm evidence can be presented to support this view. So far the most likely place in which the claim that this notion is correct seems justified is for the teleost fishes.

There are several reasons why the fishes represent a promising group in which to evaluate the relationships between evolved color vision and environmental light conditions. First, the spectral properties of photopigments have been characterized in a large number of different species of fish. Second, the photopigment systems in fish have been shown to be highly labile. As noted earlier, these pigment systems often change during the life of an individual, sometimes in direct response to changes in the photic environment. Third, many different species are believed to have good color vision, although for the vast majority this faith is without a substantiating basis. Finally, aquatic environments provide a great range of photic variation both in the quantity and the spectral quality of available light.

The wide variations in photopigment absorbance characteristics and photic environments for fish was considered by Lythgoe (1968; see also 1979). He suggested that the pigment systems must have evolved so as to provide fish with the maximal visual contrast between a target and its background. Lythgoe noted that the most efficient arrangement for detecting a dark target (that is, low reflectance target against a brighter background) is one where the absorbance peak of the photopigment matches the spectral peak of the transmission of the water. On the other hand, for detecting a bright target against a darker background, the optimal arrangement is a photopigment whose spectral peak is offset from the location of the peak transmission of the medium. Two qualifications should be mentioned; Lythgoe (1968) was concerned with scotopic vision only and, furthermore, the "contrast" of a target was computed by considering the spectral properties of the background light, the target, and the visual pigment. That is, no implications were drawn as to how all this information is processed by the fish visual system, so in point of fact what was computed was only a potential visual contrast.

In an extensive research program, Munz and McFarland (1977) have expanded Lythgoe's idea to include photopic vision, and have additionally provided a broad base of empirical measurements, including (1) measurement of the spectral properties of the photopigments present in the retinas of a large number of teleost species and (2) detailed measurements of the photic environments in which these species live. The correlation between the photopigment properties and the photic environments is sufficiently compelling to lead them to conclude, as Lythgoe suggested, that maximal visual contrast can be provided by two types of arrangements—photopigments whose peaks are matched to or offset from the spectral transmission peaks of the water depending, respectively, on whether the target is darker or lighter than the background. Because most fish must be able to accomplish both types of discriminations then, as Munz and McFarland argue, this provides an environmental pressure to support a retina containing two classes of photopigments. Because at least two classes of photopigments are a necessary precondition for color vision, this may provide an explanatory basis for the evolution of color vision.

The major ingredient missing from this effort to link photic environment and vision is any direct knowledge about the color vision of the species involved. At this stage it should hardly be necessary to point out that there is much more to understanding color vision than the characterization of the spectral properties of cone photopigments. Nevertheless, this ecological approach to the study of comparative vision is extremely promising, and we may expect that its pursuit will yield solid information about the environmental circumstances that underlie the evolution of a color capacity.

A somewhat similar approach was utilized for a terrestrial vertebrate. Under the assumption that a major purpose of color vision must be to aid in the detection of food sources, Snodderly (1978) measured the spectral reflectance properties of fruits that comprise a principal component of the diet of a South American primate, the Titi monkey (*Callicebus torquatus*). These monkeys live in an arboreal, primarily green, habitat, and direct light measurements suggest that the principal color discriminations that they might be required to make are "among subtle shades of green or between contrasting colors and green" (Snodderly, 1978). Again, unfortunately, nothing is known about the actual color vision of *Callicebus*. However, as was noted in Chapter 5, there is good reason to believe that many South American primates may show diminished color vision in the long-wavelength portion of the spectrum. Because the measurements that Snodderly made would suggest that fine discriminations are not required in that part of the spectrum in the normal environment of this animal, there is at least a suggestive link between the evolved color system and the photic environment of the animal. Here too considerably more information will be required to determine just how tight this linkage actually is.

III. Evolutionary Steps to Color Vision

The title of this section implies a promise that cannot be fulfilled. There are, quite simply, no facts that would permit an actual recounting of the ancestral stages through which the color vision of contemporary species has arisen. What can be offered instead are some suggestions as to the kinds of adaptations that may have permitted the change from monochromatic vision to color vision. How and when these adaptational changes occurred remains for the present unknown.

A speculative scheme detailing the evolution of present-day photoreceptors has been offered by Eakin (1965). He postulates that the first photoreceptor was a photosynthetic apparatus. At an early stage this photoreceptor became associated with an organelle for locomotion. The protozoan flagellate *Euglena* is often cited as a contemporary representative of this stage (Wolken, 1965). From these early photoreceptors Eakin (1965) traces two lines of descent, one involving the ciliary-based photoreceptors characteristic of the vertebrates and the other involving the rhabdomeric photoreceptors that are characteristic of present-day invertebrates.

Color vision is frequently argued to have evolved independently in several different groups of vertebrates. Where in the long history of the vertebrates does color vision first appear? Only a few estimates have been offered in answer to this question. For example, McFarland and Munz (1975) believe that the presence of multiple photopigments first occurred during the Silurian period when vertebrates changed from bottom-dwelling to free-swimming. They reason that the free swimmers would require higher photocontrasts to provide for efficient food localization in their new environment, and that this need may have provided the necessary pressure to maintain the presence of multiple retinal photopigments, a condition they appear to equate with color vision. Stephenson (1973) has specifically considered the evolution of color vision among the primates. He concludes that color vision in this order may date to the early Tertiary or the late Cretaceous period—at least 70 million years ago. This conclusion is based on the dating of the only known fossil primate common in distribution to both Old and New Worlds. Such a connection would appear to assume that the evolutionary determinants of color vision have been the same in Old and New World primates. That proposition is, at the least, highly debatable.

If one considers eyes with a single photopigment as the starting point and assumes that the color vision found in some contemporary vertebrates (for exam-

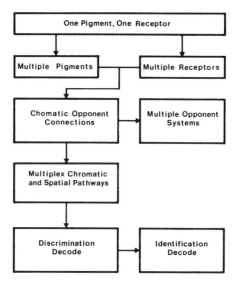

Figure 6.4   This flow chart suggests the kinds of alterations that may have occurred in the course of the evolution of color vision systems. Each stage is discussed in the text.

ple, our own species) is a representative end-product, then several changes must have occurred. Figures 6.4 illustrates in the form of a flow chart what some of these steps might reasonably have been. Some of these steps have also been suggested by others (Hemila, Reuter, & Virtanen, 1976). A few amplifying comments can be offered on each of these possible steps.

A. One Pigment and One Receptor Type

Although, as noted previously, this situation likely characterized the most primitive photoreceptors, it is an arrangement that was apparently abandoned by those vertebrates to whom vision assumed a predominant role as a means of sensory access to the environment. There may well be some contemporary vertebrates whose eyes contain only a single photoreceptor type and a single class of photopigment, who are truly and consistently monochromatic (Munz & McFarland, 1977). If so, they are very uncommon. Even those contemporary species that are claimed to come close to this status often give a hint that their retinas do not really have a single type of receptor and a single class of photopigment. For example, although the skate (*Raja oscellata*) seems to have only rods and a single type of photopigment, there are nevertheless some functional aspects to its vision (specifically in its ability to resolve rapid temporal changes) that suggest that it may not fill the bill as a single-pigment, single-receptor-type eye (Green & Siegel, 1975). At any rate, although this situation was undoubtedly the starting point for the evolution of color vision systems, most contemporary species have moved well beyond its highly restricted visual confines.

B. Multiple Photopigments

As the existence of visual pigment nomograms indicates, photopigments show only modest variations in their spectral bandwidths. This being the case, the only means a species has of broadening its spectral sensitivity, of opening its spectral window, is to acquire multiple photopigments with different λ_{max} values. Whether or not color vision follows from this arrangement is a quite separate question; the adaptation may arise purely from a need to deal with new spectral energy distributions. A number of contemporary species may be cited as providing examples of this kind of adaptation.

One instance of this sort of arrangement occurs in the American eel (*Anguilla*) whose retina contains two rhodopsins with λ_{max} of 482 and 501 nm, and a porphyropsin with a λ_{max} of 523 nm. As these eels migrate to the sea and undergo metamorphosis, the photopigment complement of the retina changes from one containing a mixture of the 523- and 501-nm pigments to one containing principally the 482- and 501-nm pigments (Beatty, 1975). This change in retinal photopigment yields an increase in spectral sensivity to the short wavelengths, which is argued to parallel the eels' entry into a marine environment where the photic surroundings are much richer in short-wavelength light.

A different and interesting example of this same kind of adaptation is seen in the minnow (*Notemigonus crysoleucas*) whose retina contains two photopigments, a rhodopsin with λ_{max} of 502 nm and a porphyrhopsin with a λ_{max} of 529 nm (Bridges, 1964). In studying the relative proportions of these two pigments in the eyes of a number of individuals, Bridges found that the proportions of the two pigments varied by as much as 44% in different eyes. Variations in these proportions would, of course, produce variations in the spectral sensitivities of the fish. Because this minnow species lives in large schools in which there are very tight behavioral relationships between the members of the school, Bridges (1964) suggests that the individual variation in pigment proportions might be seen as a means of broadening the spectral sensitivity of the entire group, and thereby achieving something quite beyond what could be obtained if all these fish had the same photopigments in the same proportions.

C. Multiple Receptor Types

Starting from a prototypical eye that contained one receptor class and one type of photopigment, it is quite possible that an initial step involved the addition of a second class of photoreceptors. This might occur, for example, if it became advantageous to increase the range of photic intensity over which the eye needed to be able to function. Such an increase might arise by virtue of the fact that the new photoreceptor differed in structure, perhaps because it funneled quanta more efficiently into the outer segment. Or the new photoreceptor might have increased the sensitivity range of the visual system because of the kinds of connections it made with cells located more distally.

This kind of adaptational step could have occurred quite independently of any changes in the character of the retinal photopigment. Munz and McFarland (1977) cite the skipjack tuna (*Katsuwonus pelamis*) as an example of a vertebrate whose retina contains two classes of photoreceptors (rods and cones) but only a single class of retinal photopigment (a rhodopsin with a λ_{max} of 483 nm). In this case it is not known what visual advantage there is to having two classes of receptors containing the same photopigment, but it could well be along the lines suggested in the previous paragraph.

It is possible that the coexistence of single and double cones in the same retina constitutes another example of this same process. As an example, it has been established that in some teleost fish the relative proportions of single and double cones changes when their habitat changes. Thus, as the rockfish (*Sebastes diploproa*) migrates from its juvenile home at the sea surface to depths of several hundred meters where it spends its adult life, there is an increase in the proportion of double cones. This is apparently brought about by the fusion of single cones (Boehlert, 1978). It can be speculated that this change yields an increase in visual sensitivity at the cost of some reduction in visual resolution. Although this example is in a direction opposite to that required to move toward a competent color vision system, it indicates that receptor transmutation can yield visual benefits without (apparently) any obviously necessary change in the number or type of photopigments available.

The fact that one can list contemporary representatives of these two initial directions of change from the one pigment/one receptor prototype suggests that they might represent quite separate steps in the evolution of color vision. On the other hand, these may be changes that occur contemporaneously. Whatever the case, those groups that achieved multiple photopigment classes and multiple receptor types were well on the path to the retinal organization found in all contemporary vertebrates that have color vision.

D. Evolutionary Sequence of Cone Types

Given that multiple classes of cones, each containing a photopigment having a different λ_{max} value, are the end-products of an evolved color vision system, it is quite conceivable that these multiple classes were elaborated in a sequential series, rather than simultaneously. The particular ordering of their acquisition undoubtedly varies among different animal groups depending on the particular environmental constraints operative at the time.

Although speculative in the extreme, it is possible to offer the following scenario for the sequence of mammalian cone types. The primary purpose of cone vision is to permit the higher spatial and temporal resolution so useful in bright light environments. The cones responsible for these functions in mammals are those having λ_{max} values in the middle wavelength range. Among such cone classes, at least among the primates, there are few significant differences in their retinal distributions, their structure, or in their spatial and temporal response properties. To achieve the principal goal of cone vision, therefore, a first step would involve the acquisition of a photopigment class behaving like those in mammals, which have a λ_{max} somewhere in the middle wavelength range. Precisely where its spectral peak might be would obviously depend on the photic environment in which it evolved, but let us assume that like the present mammalian models, it would have its spectral peak in the middle wavelengths. Once this step was achieved, the advantages offered by color vision could be added by the acquisition of a second cone class (along with the appropriate neural differencing networks). What might this second cone class be? What is so far clearly known is that among contemporary mammals there are no tritanopic species. Consequently, one might suppose that the second cone class would be one having a λ_{max} in the short wavelengths. With these two cone classes in place the animal would have the advantages of the high spatial and temporal resolution provided by the middle wavelength cones along with the color vision capacity provided by the addition of a short-wavelength photopigment. The resulting dichromacy would permit coarse color vision over a wide portion of the spectrum, in particular the ability to discriminate "blues" from other stimuli occupying the spectral ranges beyond about 500 nm. The need for more subtle color discriminations among the longer-wavelength stimuli could then be accommodated by the acquisition of a second cone class having maximum sensitivity in the middle to long wavelengths.

The several unique properties of the short-wavelength cones mentioned earlier invite the further speculation that these cones are more closely allied to retinal rods than to the long-wavelength cones. The formation of new middle-to-long wavelength cone classes, on the other hand, can be supposed to arise from selective pressures operating on the normal variations in λ_{max} values that appear to characterize these cone classes, a fact also noted previously.

E. Chromatically Opponent Connections

Given the relatively limited number of vertebrate species from whom information is available, it is probably unwise to draw any general conclusions about color vision mechanisms. Nevertheless, it is becoming increasingly more evident that cells showing chromatically opponent response patterns are a widespread, if not universal, feature of all visual systems known to yield color vision. If alternative schemes for coding color information have been exploited, there do not seem to be any contemporary representatives in which they might be convincingly illustrated. One must suppose that an early step in the evolution of color vision systems involved the development of neural organizations to make antagonistic comparisons of the signals generated by photoreceptors having different spectral sensitivities.

In the context of how other information is processed by the visual system, it is hardly surprising that color information is transmitted in the form of opponent interactions. Antagonistic interactions constitute the principal means for processing visual information in general whether it be spatial, temporal, or chromatic. Because the processing of spatial information by visual systems utilizes opponent comparisons, and since the analysis of spatial features is the primary function of visual systems, it can be guessed that the use of similar mechanisms to handle color information represents a quite logical evolutionary change. The primary utility of opponent-type interactions is that small differences within the stimulus array, whether they be spatial, temporal, or chromatic in nature can be highlighted by nervous system devices that compute differences. Such small differences are a feature of color vision systems that are in general based on the outputs from photopigments whose absorption spectra are both broad and generally overlapping.

How many such opponent systems evolve is obviously limited by the number of photopigment systems available. Beyond that, as was noted in Chapter 3, some opponent combinations are apparently much more likely to be made than others. Presumably the rules governing which opponent connections are preferred ones are set by some of the considerations raised in the previous sections.

F. Common Neural Channels for Spatial and Chromatic Information

Because color is a property of perceived objects, the processing of spatial and chromatic information must proceed jointly at some level in the visual system. Although it is possible to imagine a complete independence of transmission channels for color and space from the eye to some central nervous system location, this is quite clearly not an arrangement used in any vertebrate visual system that has been studied. Indeed, analysis of receptive field organizations makes it clear that these aspects of the visual environment are typically processed and transmitted along the same pathways. Because color is used as one means of making spatial discriminations this is a sensible arrangement. Beyond that, joint transmission of color and spatial information may be a requirement imposed by the structural organization of the visual system. Although references to the optic nerve as a ''bottleneck'' through which visual information must pass to brain are frequently made in the literature, it is nevertheless cogent to repeat that there are many more visual cells on both sides of this bottleneck, and thus it may simply be a requirement based on economy that dictates that spatial and chromatic information are neurally intertwined through the afferent visual system. Speculations as to how and when this association arose in the course of visual evolution are not presently fruitful.

G. Decoding

At some nervous system location(s) the information transmitted in the form of chromatically opponent interactions must be evaluated so that the appropriate responses can be generated. The mechanisms and locations for this decoding process remain to be discovered. The two stages of this process indicated by the boxes labeled ''discrimination'' and ''identification'' in Figure 6.4 emphasize the belief documented earlier that these are both products of the visual system with the latter dependent on, but separable from, the former. Because the two are separable they could quite conceivably have arisen as sequential stages in the evolution of color vision.

References

Abramov, I. Further analysis of the responses of LGN cells. *Journal of the Optical Society of America,* **58,** 1968, 574-579.

Abramov, I. Retinal mechanisms of colour vision. In M. G. F. Fuortes (Ed.), *Handbook of sensory physiology,* Vol. VII/2, Physiology of photoreceptor organs. Berlin and New York: Springer-Verlag, 1970. Pp. 567-607.

Aguilar, M., & Stiles, W. S. Saturation of the rod mechanism of the retina at high levels of stimulation. *Optica Acta,* **1,** 1954, 59-65.

Alpern, M. What is it that confines in a world without color? *Investigative Ophthalmology,* **13,** 1974, 648-674.

Alpern, M., & Pugh, E. N. Jr. Variation in the action spectrum of erthyrolabe among deuteranopes. *Journal of Physiology,* **266,** 1977, 613-646.

Anderson, D. H., & Fisher, S. K. The photoreceptors of diurnal squirrels: Outer segment structure, disc shedding, and protein renewal. *Journal of Ultrastructure Research,* **55,** 1976, 119-141.

Anderson, D. H., Fisher, S. K., & Steinberg, R. H. Mammalian cones: Disc shedding, phagocytosis, and renewal. *Investigative Ophthalmology and Visual Science,* **17,** 1978, 117-133.

Anderson, D. H., & Jacobs, G. H. Color vision and visual sensitivity in the California ground squirrel, *Citellus beecheyi. Vision Research,* **12,** 1972, 1995-2004.

Anderson, K. V., & Symmes, D. The superior colliculus and higher visual functions in the monkey. *Brain Research,* **13,** 1969, 37-52.

Ariga, S., Dukelow, W. R., Emley, G. S., & Hutchinson, R. R. (1978). Possible errors in identification of squirrel monkeys (*Saimiri sciureus*) from different South American points of export. *Journal of Medical Primatology,* **7,** 1978, 129-135.

Autrum, H., & Thomas, I. Comparative physiology of colour vision in animals. In R. Jung (Ed.), *Handbook of sensory physiology,* Vol. VII/3A, Central visual information. Berlin and New York: Springer-Verlag, 1973. Pp. 661-692.

Backhaus, D. Experimentalle Untersuchungen uber die Sehscharfe und das Farbsehen einiger Huftiere. *Zeitschrift Fur Tierpsychologie,* **16,** 1959, 445-468. (a)

Backhaus, D. Experimentalle Prufung des Farbsehvermogens einer Massai-Giraffe (*Giraffa camelopardalis tippelskirchi* Matschie 1898). *Zeitschrift Fur Tierpsychologie,* **16,** 1959, 468-477. (b)

Backstrom, A. C., & Reuter, T. Receptive field organization of ganglion cells in the frog retina: Contributions from cones, green rods and red rods. *Journal of Physiology,* **246,** 1975, 79-107.

Baylor, D. A. Lateral interaction between vertebrate photoreceptors. *Federation Proceedings,* **33,** 1974, 1074-1077.

Baylor, D. A., Fuortes, M. G. F., & O'Bryan, D. M. Receptive fields of cones in the retina of the turtle. *Journal of Physiology,* **214,** 1971, 265-294.

Baylor, D. A., & Hodgkin, A. L. Detection and resolution of visual stimuli by turtle photoreceptors. *Journal of Physiology,* **234,** 1973, 163-198.

Barlow, H. B., & Levick, W. R. Changes in the maintained discharge rate with adaptation level in the cat retina. *Journal of Physiology,* **202,** 1969, 699–718.

Beatty, D. D. Visual pigments of the American eel *Anguilla rostrata. Vision Research,* **15,** 1975, 771–776.

Beauchamp, R. D., & Rowe, J. S. Goldfish spectral sensitivity: A conditioned heart rate measure in restrained or curarized fish. *Vision Research,* **17,** 1977, 617–624.

Birch, D., & Jacobs, G. H. Behavioral measurements of rat spectral sensitivity. *Vision Research,* **15,** 1975, 687–691.

Birjkow, G. Vergleichende Untersuchungen uber das Helligkeits und Farbensehen bei Amphibien. *Zeitschrift Vergleichende Physiologie,* **66,** 1950, 129–178.

Blakeslee, B., & Jacobs, G. H. Color vision in the spider monkey (*Ateles*). *Folia Primatologica,* 1981, in press.

Bloch, S., & Martinoya, C. Are colour oil droplets the basis of the pigeon's chromatic space? *Vision Research Supplement 3,* 1971, 411–418.

Blough, D. S. Spectral sensitivity in the pigeon. *Journal of the Optical Society of America,* **47,** 1957, 827–833.

Blough, D. S., & Yager, D. Visual psychophysics in animals. In D. Jameson & L. M. Hurvich (Eds.), *Handbook of sensory physiology,* Vol. VII/4, *Visual psychophysics.* Berlin and New York: Springer-Verlag, 1972. Pp. 732–763.

Blough, P. M. Wavelength generalization and discrimination in the pigeon. *Perception and Psychophysics,* **12,** 1972, 342–348.

Blough, P. M. The pigeon's perception of saturation. *Journal of the Experimental Analysis of Behavior,* **24,** 1975, 135–148.

Boehlert, G. W. Intraspecific evidence for function of single and double cones in the teleost retina. *Science,* **202,** 1978, 309–311.

Bowmaker, J. K. The visual pigments, oil droplets and spectral sensitivity of the pigeon. *Vision Research,* **17,** 1977, 1129–1138.

Bowmaker, J. K. Colour vision in birds and the role of oil droplets. *Trends in Neuroscience,* August, 1980, 196–199.

Bowmaker, J. K., & Dartnall, H. J. A. Visual pigments of rods and cones in a human retina. *Journal of Physiology,* **298,** 1980, 501–511.

Bowmaker, J. K., Dartnall, H. J. A., Lythgoe, J. N., & Mollon, J. D. The visual pigments of rods and cones in the rhesus monkey (*Macaca mulatta*). *Journal of Physiology,* **274,** 1978, 329–348.

Bowmaker, J. K., Dartnall, H. J. A., & Mollon, J. D. The violet-sensitive receptors of primate retinae. *Investigative Ophthalmology and Visual Science Supplement,* **18,** 1979, 31.

Bowmaker, J. K., Dartnall, H. J. A., & Mollon, J. D. Microspectrophotometric demonstration of four classes of photoreceptors in an old world primate, *Macaca fascicularis. Journal of Physiology,* **298,** 1980, 131–143.

Bowmaker, J. K., Loew, E. R., & Liebman, P. A. Variation in the λ_{max} of rhodopsin from individual rods. *Vision Research,* **15,** 1975, 977–1003.

Boynton, R. M. Implications of the minimally-distinct border. *Journal of the Optical Society of America,* **63,** 1973, 1037–1043.

Boynton, R. M. *Human color vision.* New York: Holt, 1979.

Boynton, R. M., & Whitten, D. W. Selective adaptation in primate photoreceptors. *Vision Research,* **12,** 1972, 855–874.

Bridges, C. D. B. Visual pigments of the pigeon (*Columba livia*). *Vision Research,* **2,** 1962, 125–137.

Bridges, C. D. B. Variation of visual pigment amongst individuals of an American minnow, *Notemigonus crysoleucas boscii. Vision Research,* **4,** 1964, 233–240.

Bridges, C. D. B. Spectroscopic properties of porphyropsins. *Vision Research,* **7,** 1967, 349–367.

Bridges, C. D. B. The rhodopsin-porphyropsin visual system. In H. J. A. Dartnall (Ed.), *Handbook of sensory physiology,* Vol. VII/1, Photochemistry of vision. Berlin and New York: Springer-Verlag, 1972. Pp. 417–480.

Brown, J. L., Shively, F. D., LaMotte, R. H., & Sechzer, J. A. Color discrimination in the cat. *Journal of Comparative and Physiological Psychology,* **84,** 1973, 534–544.

Buchenauer, D., & Fritsch, B. Colour vision in domestic goats *(Capra hircus L.). Zeitschrift für Tierpsychologie,* **53,** 1980, 225–230.

Burnham, R. W., Hanes, R. M., & Bartleson, C. J. *Color: A Guide to Basic Facts and Concepts.* New York: Wiley, 1963.

Chapman, R. M. Light wavelength and energy preference of the bullfrog: Evidence for color vision. *Journal of Comparative and Physiological Psychology,* **61,** 1966, 429–435.

Christ, R. E. Review and analysis of color coding research for visual displays. *Human Factors,* **17,** 1975, 542–570.

Cicerone, C. M. Cones survive rods in the light-damaged eye of the albino rat. *Science,* **194,** 1976, 1183–1185.

Clayton, K. N., & Kamback, M. Successful performance by cats on several colour discrimination problems. *Canadian Journal of Psychology,* **20,** 1966, 173–182.

Clark, W. E. LeGros *Anatomical patterns as the essential basis of sensory discrimination.* London and New York: Oxford University Press (Blackwell), 1947.

Cohen, A. I. Some observations of the fine structure of the retinal receptors of the American grey squirrel. *Investigative Ophthalmology,* **3,** 1964, 198–216.

Cohen, A. I. Rods and cones. In M. G. F. Fuortes (Ed.), *Handbook of sensory physiology,* Vol. VII/2, Physiology of photoreceptor organs. Berlin and New York: Springer-Verlag, 1972. Pp. 63–110.

Coleman, T. B., & Hamilton, W. F. Color blindness in the rat. *Journal of Comparative Psychology,* **15,** 1933, 171–181.

Cooper, G. F., & Robson, J. G. The yellow colour of the lens of the grey squirrel *(Sciurus carolinensis leucotis). Journal of Physiology,* **203,** 1969, 403–410.

Craft, L. H., & Butter, C. M. Effects of striate cortex removal on wavelength discrimination in rats. *Psychological Record,* **18,** 1968, 311–316.

Crescitelli, F. The visual cells and visual pigments of the vertebrate eye. In H. J. A. Dartnall (Ed.), *Handbook of sensory physiology,* Vol. VII/1, *Photochemistry of vision.* Berlin and New York: Springer-Verlag, 1972. Pp. 245–263.

Crescitelli, F., & Pollack, J. D. Dichromacy in the antelope ground squirrel. *Vision Research,* **12,** 1972, 1553–1586.

Darbrowska, B., Harmata, W., Lenkiewicz, Z., Schiffer, Z., & Wojtusiak, R. J. Colour perception in cows. *Behavioural Processes,* **6,** 1981, 1–10.

Dartnall, H. J. A. The interpretation of spectral sensitivity curves. *British Medical Bulletin,* **9,** 1953, 24–30.

Dartnall, H. J. A. Visual pigment from a pure cone retina. *Nature (London),* **188,** 1960, 475–479.

Dartnall, H. J. A. The identity and distribution of visual pigments in the animal kingdom. In H. Davson (Ed.), *The eye,* Vol. 2. New York: Academic Press, 1962. Pp. 367–426.

Dartnall, H. J. A. *Handbook of sensory physiology,* Vol. VII/1, Photochemistry of vision. Berlin and New York: Springer-Verlag, 1972.

Dartnall, H. J. A., Arden, G. B., Ikeda, H., Luck, C. P., Rosenberg, C. M., Pedler, H., & Tansley, K. Anatomical, electrophysiological and pigmentary aspects of vision in the bush baby: An interpretive study. *Vision Research,* **5,** 1965, 399–424.

Davidoff, J. Hemispheric sensitivity differences in the perception of colour. *Quarterly Journal of Experimental Psychology,* **28,** 1976, 387–394.

Daw, N. W. Neurophysiology of color vision. *Physiological Review, 53,* 1973, 571–611.

Daw, N. W., & Pearlman, A. L. Cat colour vision: One cone process or several. *Journal of Physiology, 201,* 1969, 745–764.

Dean, P. Visual cortex ablation and thresholds for successively presented stimuli in rhesus monkeys: II. Hue. *Experimental Brain Research, 35,* 1979, 69–83.

De Monasterio, F. M. Asymmetry of on- and off-pathways of blue-sensitive cones of the retina of macaques. *Brain Research, 166,* 1979, 39–48.

De Monasterio, F. M., & Gouras, P. Functional properties of ganglion cells of the rhesus monkey retina. *Journal of Physiology, 151,* 1975, 167–195.

De Monasterio, F. M., Gouras, P., & Tolhurst, D. J. Trichromatic colour opponency in ganglion cells of the rhesus monkey retina. *Journal of Physiology, 251,* 1975, 197–216.

De Monasterio, F. M., & Schein, S. J. Protan-like spectral sensitivity of foveal Y ganglion cells of the retina of macaque monkeys. *Journal of Physiology, 299,* 1980, 385–396.

De Valois, R. L. Analysis and coding of color vision in the primate visual system. *Cold Spring Harbor Symposium on Quantitative Biology, 30,* 1965, 567–579.

De Valois, R. L. Central mechanisms of color vision. In R. Jung (Ed.), *Handbook of sensory physiology,* Vol. VII/3 Part A, Central visual information. Berlin and New York: Springer-Verlag, 1973. Pp. 209–253.

De Valois, R. L., Abramov, I., & Jacobs, G. H. Analysis of response patterns of LGN cells. *Journal of the Optical Society of America, 56,* 1966, 966–977.

De Valois, R. L., & De Valois, K. K. Neural coding of color. In E. C. Carterette & M. P. Friedman (Eds.), *Handbook of perception,* Vol. V. New York: Academic Press, 1975. Pp. 117–166.

De Valois, R. L., & De Valois, K. K. Spatial vision. *Annual Review of Psychology, 31,* 1980, 309–341.

De Valois, R. L., & Jacobs, G. H. Vision. In A. M. Schrier and F. Stollnitz (Eds.), *Behavior of nonhuman primates,* Vol. 3. New York: Academic Press, 1971. Pp. 107–157.

De Valois, R. L., & Jacobs, G. H. Neural mechanisms of color vision. In I. Darian-Smith (Ed.), *Handbook of physiology,* Vol. III. Sensory Processes. Washington D.C.: American Physiological Society, in press.

De Valois, R. L., & Morgan, H. C. Psychophysical studies of monkey vision. II. Squirrel monkey wavelength and saturation discrimination. *Vision Research, 14,* 1974, 69–73.

De Valois, R. L., Morgan, H. C., Polson, M. C., Mead, W. R., and Hull, E. M. Psychophysical studies of monkey vision. I. Macaque luminosity and color vision tests. *Vision Research, 14,* 1974, 53–67.

De Valois, R. L., Smith, C. J., Karoly, A. J., and Kitai, S. T. Electrical responses of primate visual system. I. Different layers of macaque lateral geniculate nucleus. *Journal of Comparative and Physiological Psychology, 51,* 1958, 662–668.

Dodwell, P. C. Contemporary theoretical problems in seeing. In E. C. Carterette and M. P. Friedman (Eds.), *Handbook of perception.* Vol. V. New York: Academic Press, 1975. Pp. 57–77.

Dow, B. M. Functional classes of cells and their laminar distribution in monkey visual cortex. *Journal of Neurophysiology, 37,* 1974, 927–946.

Dowling, J. E. Structure and function in the all-cone retina of the ground squirrel. In *Physiological basis for form discrimination.* Providence, Rhode Island: Brown University, 1964. Pp. 17–23.

Dowling, J. E. Visual adaptation: Its mechanism. *Science, 157,* 1967, 584–585.

Dowling, J. E. Synaptic arrangements in the vertebrate retina: The photoreceptor synapse. In M. V. L. Bennett (Ed.), *Synaptic transmission and neuronal integration.* New York: Raven Press, 1974. Pp 87–103.

Dowling, J. E., and Boycott, B. B. Organization of the primate retina: Electron microscopy. *Proceedings of the Royal Society of London, 166B,* 1966, 80–111.

Duecker, G., & Schultze, I. Color vision and color preference in Japanese quail (*Cortunix cortunix japonica*) with colorless oil droplets. *Journal of Comparative and Physiological Psychology,* **91,** 1977, 1110-1117.

Eakin, R. M. Evolution of photoreceptors. *Cold Spring Harbor Symposium on Quantitative Biology,* **30,** 1965, 363-370.

Ebrey, T. G., & Honig, B. New wavelength dependent visual pigment nomograms. *Vision Research,* **17,** 1977, 147-151.

Ehrlich, A., & Calvin, W. H. Visual discrimination behavior in galago and owl monkey. *Psychonomic Science,* **9,** 1967, 509-510.

Emmerton, J., & Delius, J. D. Wavelength discrimination in the "visible" and ultraviolet spectrum by pigeons. *Journal of Comparative Physiology,* **141,** 1980, 47-52.

Enoch, J. M. Validation of an indicator of mammalian retinal receptor response. Action spectrum. *Journal of the Optical Society of America,* **54,** 1964, 368-374.

Enroth-Cugell, C., Hertz, B. G., & Lennie, P. Convergence of rod and cone signals in the cat's retina. *Journal of Physiology,* **269,** 1977, 297-318.

Fain, G. L. Quantum sensitivity of rods in the toad retina. *Science,* **187,** 1975, 838-841.

Fain, G. L., & Dowling, J. E. Intracellular recordings from single rods and cones in the mudpuppy retina. *Science,* **180,** 1973, 1178-1180.

Fantz, R., Ordy, J., & Udelf, M. Maturation of pattern vision in infants during the first 6 months of life. *Journal of Comparative and Physiological Psychology,* **55,** 1962, 907-917.

Fite, K. V., Soukup, J., & Carey, R. G. Wavelength discrimination in the leopard frog: A reexamination. *Brain, Behavior, & Evolution,* **15,** 1978, 405-414.

Friedman, H. Colour vision in the Virginia oppossum. *Nature (London),* **213,** 1967, 835-836.

Fuortes, M. G. F., Schwartz, E. A., & Simon, E. J. Colour-dependence of cone responses in the turtle retina. *Journal of Physiology,* **234,** 1973, 199-216.

Fuortes, M. G. F., & Simon, E. J. Interactions leading to horizontal cell responses in the turtle retina. *Journal of Physiology,* **240,** 1974, 177-198.

Garner, W. R. *The processing of information and structure.* Potomac, Maryland: Lawrence Erlbaum, 1974.

Geschwind, N., & Fusillo, M., Color naming defects in association with alexia. *Archives of Neurology,* **15,** 1966, 137-146.

Ginsburg, A. P. The perception of visual form: A two dimensional filter analysis. In V. D. Glezer (Ed.), *Information processing in the visual system* (*Proceedings of the IV symposium on sensory system physiology*). Leningrad, USSR, 1976. Pp. 46-51.

Glickman, S. E., Clayton, K., Schiff, B. Guritz, D., & Messe, L. Discrimination learning in some primitive mammals. *Journal of Genetic Psychology,* **106,** 1965, 325-335.

Goldsmith, T. H. Hummingbirds see near ultraviolet light. *Science,* **207,** 1980, 786-788.

Gouras, P. Color opponency from fovea to striate cortex. *Investigative Ophthalmology,* **11,** 1972, 427-434.

Gouras, P. Opponent-colour cells in different layers of foveal striate cortex. *Journal of Physiology,* **238,** 1974, 583-602.

Gouras, P., & Kruger, J. Responses of cells in foveal visual cortex of the monkey to pure color contrast. *Journal of Neurophysiology,* **42,** 1979, 850-860.

Gouras, P., & Link, K. Rod and cone interaction in dark-adapted monkey ganglion cells. *Journal of Physiology,* **184,** 1966, 499-510.

Govardovskii, V. I., & Zueva, L. V. Visual pigments of chicken and pigeon. *Vision Research,* **17,** 1977, 537-543.

Graf, V. A spectral sensitivity curve and wavelength discrimination for the turtle (*Chrysemys picta picta*). *Vision Research,* **7,** 1967, 915-928.

Graf, V., & Norren, D. V. A blue sensitive mechanism in the pigeon retina: λ_{max}400 nm. *Vision Research,* **14,** 1974, 1203-1209.

Granda, A. M., & Dvorak, C. A. Vision in turtles. In F. Crescitelli (Ed.), *Handbook of sensory physiology,* Vol. VII/5, *The visual system in vertebrates.* Berlin and New York: Springer-Verlag, 1977. Pp. 451-495.

Granger, E. M., & Heurtley, J. C. Visual chromaticity-modulation transfer function. *Journal of the Optical Society of America,* **63,** 1973, 1173-1174.

Granit, R. *Sensory mechanisms of the retina.* London and New York: Oxford University Press, 1947.

Granit, R. *Receptors and Sensory Perception.* New Haven, Connecticut: Yale University Press, 1955.

Green, D. G., & Dowling, J. E. Electrophysiological evidence for rod-like receptors in the gray squirrel, ground squirrel and prairie dog retinas. *Journal of Comparative Neurology,* **159,** 1975, 461-472.

Green, D. G., & Siegel, I. M. Double branched flicker fusion curves from all-rod skate retina. *Science,* **188,** 1975, 1120-1122.

Grether, W. F. Color vision and color blindness in monkeys. *Comparative Psychology Monographs,* **15,** 1939, 1-38.

Grether, W. F. Chimpanzee color vision. I. Hue discrimination at 3 spectral points. *Journal of Comparative Psychology,* **29,** 1940, 167-77. (a)

Grether, W. F. Chimpanzee color vision. II. Color mixture properties. *Journal of Comparative Psychology,* **29,** 1940, 179-186. (b)

Grether, W. F. Chimpanzee color vision. III. Spectral limits. *Journal of Comparative Psychology,* **29,** 1940, 187-192. (c)

Grether, W. F. A comparison of human and chimpanzee spectral hue discrimination curves. *Journal of Experimental Psychology,* **26,** 1940d, 394-403. (d)

Grether, W. F. Spectral saturation curves for chimpanzee and man. *Journal of Experimental Psychology,* **28,** 1941, 419-427.

Grutzner, P. Acquired color vision defects. In D. Jameson and L. M. Hurvich (Eds.), *Handbook of sensory physiology,* Vol. VII/4, *Visual psychophysics.* Berlin and New York: Springer-Verlag, 1972. Pp. 643-659.

Grzimek, B. Versuche uber das Farbsehen von Pflanzenessern. *Zeitschrift fur Tierpsychologie,* **9,** 1952, 23-39.

Gunter, R. The discrimination between lights of different wavelengths in the cat. *Journal of Comparative and Physiological Psychology,* **47,** 1954, 169-172.

Gunter, R., Feigenson, L., & Blakeslee, P. Color vision in the cebus monkey. *Journal of Comparative and Physiological Psychology,* **60,** 1965, 107-113.

Gur, M., & Purple, R. L. Retinal ganglion cell activity in the ground squirrel under halothane anesthesia. *Vision Research,* **18,** 1978, 1-14.

Gur, M., & Purple, R. L. Some temporal output properties of color opponent units in the ground squirrel retina. *Brain Research,* **166,** 1979, 233-244.

Guth, S. L., Donley, N. J., & Marrocco, R. T. On luminance additivity and related topics. *Vision Research,* **9,** 1969, 537-575.

Hailman, J. P. *Optical signals.* Bloomington, Indiana: Indiana University Press, 1977.

Hailman, J. P., & Jaeger, R. G. Phototactic responses to spectrally dominant stimuli and the use of colour vision by adult anuran amphibians: A comparative study. *Animal Behaviour,* **22,** 1974, 757-795.

Hamilton, W. R., & Coleman, T. B. Trichromatic vision in the pigeon as illustrated by the spectral hue discrimination curve. *Journal of Comparative Psychology,* **15,** 1933, 183-191.

Hammond, P. The neural basis for colour discrimination in the domestic cat. *Vision Research,* **18,** 1978, 233–235.

Harosi, F. I. Spectral relations of cone pigments in goldfish. *Journal of General Physiology,* **68,** 1976, 65–80.

Helson, H. Studies of anomalous contrast and assimilation. *Journal of the Optical Society of America,* **53,** 1963, 179–184.

Hemila, S. T., Reuter, T., & Virtanen, K. The evolution of colour-opponent neurons and color vision. *Vision Research,* **16,** 1976, 1359–1362.

Hodos, W. Color discrimination deficits after lesions of the nucleus rotundus in pigeons. *Brain, Behavior, & Evolution,* **2,** 1969, 185–200.

Hoffman, G. Untersuchungen uber das Farbsehvermogen des Zebu. *Zeitschrift fur Tierpsychologie,* **9,** 1952, 170–179.

Hubel, D. H., & Wiesel, T. N. Receptive fields and functional architecture of monkey striate cortex. *Journal of Physiology,* **195,** 1968, 215–243.

Humphrey, N. K. What the frog's eye tells the monkey's brain. *Brain, Behavior, & Evolution,* **3,** 1970, 324–337.

Humphrey, N. Colour and brightness preferences in monkeys. *Nature (London),* **229,** 1971, 615–617.

Humphrey, N. "Interest" and "Pleasure": Two determinants of a monkey's visual preferences. *Perception,* **1,** 1972, 395–416.

Hurvich, L. M. Color vision deficiencies. In D. Jameson and L. M. Hurvich (Eds.), *Handbook of sensory physiology,* Vol. VII/4, *Visual psychophysics.* Berlin and New York: Springer-Verlag, 1972. Pp. 582–624.

Hurvich, L. M., & Jameson, D. *The perception of brightness and darkness.* Boston, Massachusetts: Allyn and Bacon, 1966.

Hurvich, L. M., & Jameson, D. On the measurement of dichromatic neutral points. *Acta Chromatica,* **2,** 1974, 207–216.

Jacobs, G. H. Spectral sensitivity and color vision of the squirrel monkey. *Journal of Comparative and Physiological Psychology,* **56,** 1963, 616–621.

Jacobs, G. H. Increment-threshold spectral sensitivity in the squirrel monkey. *Journal of Comparative and Physiological Psychology,* **79,** 1972, 425–431.

Jacobs, G. H. Scotopic and photopic visual capacities of an arboreal squirrel (*Sciurus niger*). *Brain, Behavior, & Evolution,* **10,** 1974, 307–321.

Jacobs, G. H. Wavelength discrimination in gray squirrels. *Vision Research,* **16,** 1976, 325–327.

Jacobs, G. H. Visual capacities of the owl monkey (*Aotus trivirgatus*): I. Spectral sensitivity and color vision. *Vision Research,* **17,** 1977, 811–820. (a)

Jacobs, G. H. Visual sensitivity: Significant within-species variations in a nonhuman primate. *Science,* **197,** 1977, 499–500. (b)

Jacobs, G. H. Spectral sensitivity and colour vision in the ground-dwelling sciurids: Results from golden-mantled ground squirrels and comparisons for five species. *Animal Behaviour,* **26,** 1978, 409–421.

Jacobs, G. H., Blakeslee, B., & Tootell, R. B. H. Color discrimination tests on fibers in the ground squirrel optic nerve. *Journal of Neurophysiology,* **45,** 1981, 903–914.

Jacobs, G. H., Bowmaker, J. K., & Mollon, J. D. Colour vision in monkeys: behavioural and microspectrophotometric measurements on the same individuals. *Nature (London),* **292,** 1981, 541–543.

Jacobs, G. H., Fisher, S. K., Anderson, D. H., & Silverman, M. S. Scotopic and photopic vision in the California ground squirrel: Physiological and anatomical evidence. *Journal of Comparative Neurology,* **165,** 1976, 209–228.

Jacobs, G. H., & Pulliam, K. A. Vision in the prairie dog: Spectral sensitivity and color vision. *Journal of Comparative and Physiological Psychology*, **84**, 1973, 240-245.

Jacobs, G. H., & Tootell, R. B. H. Spectrally-opponent responses in ground squirrel optic nerve. *Vision Research*, **20**, 1980, 9-13.

Jacobs, G. H., & Tootell, R. B. H. Spectral response properties of optic nerve fibers in the ground squirrel. *Journal of Neurophysiology*, **45**, 1981, 891-902.

Jacobs, G. H., Tootell, R. B. H., Fisher, S. K., & Anderson, D. H. Rod photoreceptors and scotopic vision in ground squirrels. *Journal of Comparative Neurology*, **189**, 1980, 113-125.

Jacobs, G. H., & Yolton, R. L. Visual sensitivity and color vision in ground squirrels. *Vision Research*, **11**, 1971, 511-537.

Jaeger, R. G., & Hailman, J. P. Two types of phototactic behavior in anuran amphibians. *Nature (London)*, **230**, 1971, 189-190.

Jaeger, W. Genetics of congenital colour deficiencies. In D. Jameson and L. M. Hurvich (Eds.), *Handbook of sensory physiology*, Vol. VII/4, Visual psychophysics. Berlin and New York: Springer-Verlag, 1972. Pp. 626-659.

Jitsumori, M. Anomaloscope experiment for a study of color mixture in the pigeon. *Japanese Psychological Research*, **18**, 1976, 126-135.

Kaneko, A. Receptive field organization of bipolar and amacrine cells in the goldfish retina. *Journal of Physiology*, **235**, 1973, 113-153.

Keating, E. G. Rudimentary color vision in the monkey after removal of striate and preoccipital cortex. *Brain Research*, **179**, 1979, 379-384.

Kicliter, E., & Loop, M. S. A test of wavelength discrimination. *Vision Research*, **16**, 1976, 951-956.

Kicliter, E., Loop, M. S., & Jane, J. A. Effects of posterior neocortical lesions on wavelength, light/dark and stripe orientation discrimination in ground squirrels. *Brain Research*, **122**, 1977, 15-21.

King-Smith, P. E. Absorption spectra and function of the coloured oil drops in the pigeon retina. *Vision Research*, **9**, 1969, 1391-1399.

Kinsbourne, M., & Warrington, E. K. Observations on colour agnosia. *Journal of Neurology, Neurosurgery and Psychiatry*, **27**, 1964, 269-299.

Klopfer, F. D., & Butler, R. L. Color vision in swine. *American Zoologist*, **4**, 1964, 294.

Kluver, H. Functional significance of the geniculo-striate system. *Biological Symposia*, **5**, 1942, 253-299.

Kohts, N. Recherches sur l'intelligence du chimpanze par la methode de choix d'Apres modele. *Journal de Psychologie Normale et Pathologique*, **25**, 1928, 255-275.

Kolosvary, G. A study of color vision in the mouse (*Mus musculus*) and the souslik (*Citellus citellus*). *Journal of Genetic Psychology*, **44**, 1934, 473-475.

Kovach, J. K., Wilson, G., & O'Connor, T. On the retinal mediation of genetic influences in color preferences of Japanese quail. *Journal of Comparative and Physiological Psychology*, **90**, 1976, 1144-1151.

Kozack, W. M., & Reitboeck, H. J. Color-dependent distribution of spikes in single optic tract fibers of the cat. *Vision Research*, **14**, 1974, 405-420.

Kreithen, M. L., & Eisner, T. Ultraviolet light detection by the homing pigeon. *Nature (London)*, **272**, 1978, 347-348.

Ladman, A. J. The fine structure of the rod-bipolar cell synapse in the retina of the albino rat. *Journal of Biophysical and Biochemical Cytology*, **4**, 1958, 459-465.

Land, E. H. Experiments in color vision. *Scientific American*, **200**, 1959, 84-99.

Lashley, K. S. The mechanism of vision. V. The structure and image-forming power of the rat's eye. *Journal of Comparative Psychology*, **13**, 1932, 173-200.

Lashley, K. S. The mechanism of vision. XV. Preliminary studies of the rat's capacity for detail vision. *Journal of General Psychology,* **18,** 1938, 123–193.

Last, R. J. *Anatomy of the eye and orbit.* Philadelphia, Pennsylvania: Saunders, 1961.

Laties, A. M., & Liebman, P. A. Cones of living amphibian eye: Selective staining. *Science,* **168,** 1970, 1475–1477.

LaVail, M. M. Survival of some photoreceptor cells in albino rats following long-term exposure to continuous light. *Investigative Ophthalmology,* **15,** 1976, 64–70.

Lemmon, V., & Anderson, K. V. Central neurophysiological correlates of constant light-induced retinal degeneration. *Experimental Neurology,* **63,** 1979, 50–75.

Lepore, R., Lassonde, M., Ptito, M., & Cardu, B. Spectral sensitivity in a female *Cebus griseus. Perceptual and Motor Skills,* **40,** 1975, 783–788.

Liebman, P. A. Microspectrophotometry of photoreceptors. In H. J. A. Dartnall (Ed.) *Handbook of sensory physiology,* Vol. VII/1, Photochemistry of vision. Berlin and New York: Springer-Verlag, 1972. Pp. 481–528.

Liebman, P. A., & Entine, G. Sensitive low-light-level microspectrophotometer detection of photosensitive pigments of retinal cones. *Journal of the Optical Society of America,* **54,** 1964, 1451–1459.

Liebman, P. A., & Entine, G. Visual pigments of frog and tadpole (*Rana pipiens*). *Vision Research,* **8,** 1968, 761–775.

Liebman, P. A., & Granda, A. M. Microspectrophotometric measurements of visual pigments in two species of turtle, *Pseudemys scripta* and *Chelonia mydas. Vision Research,* **11,** 1971, 105–114.

Loew, E. R. The visual pigments of the gray squirrel, *Sciurus carolinensis leucotis. Journal of Physiology,* **251,** 1975, 48–49P.

Loop, M. S., & Bruce, L. L. Cat color vision: The effect of stimulus size. *Science,* **199,** 1978, 1221–1222.

Lythgoe, J. N. Visual pigments and visual range underwater. *Vision Research,* **8,** 1968, 997–1012.

Lythgoe, J. N. The adaptation of visual pigments to the photic environment. In H. J. A. Dartnall (Ed.), *Handbook of sensory physiology,* Vol. VII/1, Photochemistry of vision. Berlin and New York: Springer-Verlag, 1972. Pp. 566–603.

Lythgoe, J. N. *The ecology of vision.* London and New York: Oxford University Press, 1979.

McCleary, R. A., & Bernstein, J. A unique method for control of brightness cues in the study of color vision in fish. *Physiological Zoology,* **32,** 1959, 284–292.

McFarland, W. N., & Munz, F. W. The evolution of photopic visual pigments in fishes. *Vision Research,* **15,** 1975, 1071–1080.

MacKay, D. M., & Jeffreys, D. A. Visually evoked potentials and visual perception in man. In R. Jung (Ed.), *Handbook of sensory physiology,* Vol. VII/3B, Central visual information. Berlin and New York: Springer-Verlag, 1973. Pp. 647–678.

Malmo, R. B., & Grether, W. F. Further evidence of red blindness (protanopia) in cebus monkeys. *Journal of Comparative and Physiological Psychology,* **40,** 1947, 143–147.

Marc, R. C., & Sperling, H. G. Chromatic organization of primate cones. *Science,* **196,** 1977, 454–456.

Marks, W. B. Visual pigments of single goldfish cones. *Journal of Physiology,* **178,** 1965, 14–32.

Marriott, F. H. C. Colour vision: Other phenomena. In H. Davson (Ed.), *The eye,* Vol. 2, *The visual process.* New York: Academic Press, 1962. Pp. 273–297.

Marrocco, R. T., & Li, R. H. Monkeys superior colliculus: Properties of single cells and their afferent inputs. *Journal of Neurophysiology,* **40,** 1977, 844–860.

Martin, G. R., & Muntz, W. R. A. Retinal oil droplets and vision in the pigeon (*Columba livia*). In A. M. Granda and J. H. Maxwell (Eds.), *Neural mechanisms of behavior in the pigeon.* New York: Plenum Press, 1979. Pp. 307–325.

Mello, N. K., & Peterson, N. J. Behavioral evidence for color discrimination in cat. *Journal of Neurophysiology,* **31,** 1964, 249–282.

Mervis, R. F. Evidence of color vision in a diurnal prosimian, *Lemur catta. Animal Learning and Behavior,* **2,** 1974, 238–240.

Meyer, D. B., Cooper, T. G., & Gernez, C. Retinal oil droplets. In J. W. Rohen (Ed.), *Structure of the eye,* Vol. 2. Berlin and New York: Springer-Verlag, 1965. Pp. 521–533.

Meyer, D. B., Stuckey, S. R., & Hudson, R. A. Oil droplet carotenoids of avian cones. I. Dietary exclusion: Models for biochemical and physiological studies. *Comparative Biochemical Physiology,* **40B,** 1971, 61–76.

Meyer, D. R., Miles, R. C., & Ratoosh, P. Absence of color vision in cat. *Journal of Neurophysiology,* **17,** 1954, 289–294.

Michael, C. R. Receptive fields of single optic nerve fibers in a mammal with an all-cone retina. *Journal of Neurophysiology,* **31,** 1968, 249–282.

Michael, C. R. Opponent-color and opponent-contrast cells in lateral geniculate nucleus of the ground squirrel. *Journal of Neurophysiology,* **35,** 1972, 536–550.

Michael, C. R. Visual receptive fields of single neurons in superior colliculus of the ground squirrel. *Journal of Neurophysiology,* 36, 1973, 815–832.

Michael, C. R. Color vision mechanisms in monkey striate cortex: Dual-opponent cells with concentric receptive fields. *Journal of Neurophysiology,* **41,** 1978, 572–588.

Michael, K. M., Fischer, B. E., & Johnson, J. I. Raccoon performance on color discrimination problems. *Journal of Comparative and Physiological Psychology,* **53,** 1960, 379–380.

Miles, R. C. Color vision in the squirrel monkey. *Journal of Comparative and Physiological Psychology,* **51,** 1958, 328–331. (a)

Miles, R. C. Color vision in the marmoset. *Journal of Comparative and Physiological Psychology,* **51,** 1958, 152–154. (b)

Miles, R. C., Ratoosh, P., & Meyer, D. R. Absence of color vision in guinea pig. *Journal of Neurophysiology,* **19,** 1956, 254–258.

Miller, W. H., & Snyder, A. W. The tiered vertebrate retina. *Vision Research,* **17,** 1977, 239–255.

Moreland, J. D., & Lythgoe, J. N. Yellow corneas in fishes. *Vision Research,* **8,** 1968, 1377–1380.

Muntz, W. R. A. Effectiveness of different colors of light in releasing positive phototactic behavior of frogs, and a possible function of the retinal projection to the diencephalon. *Journal of Neurophysiology,* **25,** 1962, 712–720.

Muntz, W. R. A. The photopositive response of the frog (*Rana pipiens*) under photopic and scotopic conditions. *Journal of Experimental Biology,* **45,** 1966, 101–111.

Muntz, W. R. A. Inert absorbing and reflecting pigments, In H. J. A. Dartnall (Ed.), *Handbook of sensory physiology,* Vol. VII/1, *Photochemistry of vision.* Berlin and New York: Springer Verlag, 1972. Pp. 530–565.

Muntz, W. R. A., & Cronly-Dillon, J. Color discrimination in goldfish. *Animal Behaviour,* **14,** 1966, 351–355.

Munz, F. W., & McFarland, W. N. Evolutionary adaptation of fishes to the photic environment. In F. Crescitelli (Ed.), *Handbook of sensory physiology,* Vol. VI/5, *The visual system in vertebrates.* Berlin and New York: Springer-Verlag, 1977. Pp. 194–274.

Munz, F. W., & Schwanzara, S. A. A nomogram for retinene$_2$-based visual pigments. *Vision Research,* **7,** 1967, 1237–1244.

Nagy, A. L. Large-field substitution Rayleigh matches of dichromats. *Journal of the Optical Society of America,* **70,** 1980, 778–784.

Nelson, R. Cat cones have rod input: A comparison of the response properties of cones and horizontal cell bodies in the retina of the cat. *Journal of Comparative Neurology,* **172,** 1977, 109–136.

Nilsson, S. E. G. An electron microscopic classification of the retinal receptors of the leopard frog. *Journal of Ultrastructure Research,* **10,** 1964, 390–416.

Normann, R. A., & Werblin, F. W. Control of retinal sensitivity. I. Light and dark adaptation of vertebrate rods and cones. *Journal of General Physiology,* **63,** 1974, 37-61.

Norren, D. V., & Vos, J. J. Spectral transmission of the human ocular media. *Vision Research,* **14,** 1974, 1237-1244.

Norren, D. V. Two short wavelength sensitive cone systems in pigeon, chicken and daw. *Vision Research,* **15,** 1975, 1164-1166.

Nuboer, J. W. F. Spectral discrimination in a rabbit. *Documenta Ophthalmologica,* **30,** 1971, 279-298.

Ogden, T. E. The receptor mosaic of *Aotus trivirgatus:* Distribution of rods and cones. *Journal of Comparative Neurology,* **163,** 1975, 193-202.

Osterholm, H. The significance of distance-receptors in the feeding behavior of the fox, *Vulpes vulpes L. Acta Zoologica Fennica,* **106,** 1964, 3-31.

Oyama, T., Furusaka, T., & Kito, T. Color vision in men and animals. *Scientific American* (Japanese edition), December, 1979, 98-110.

Padmos, P., & Norren, D. V. Cone systems interaction in single neurons of the lateral geniculate nucleus of the macaque. *Vision Research,* **15,** 1975, 617-619.

Pasik, T., & Pasik, P. The visual world of monkeys deprived of striate cortex: Effective stimulus parameters and the importance of the accessory optic system. *Vision Research Suppl. 3,* 1971, 419-435.

Pearlman, A. L., Birch, J., & Meadows, J. C. Cerebral color blindness: An acquired defect in hue discrimination. *Annals of Neurology,* **5,** 1979, 253-261.

Pearlman, A. L., & Daw, N. W. Opponent color cells in the cat lateral geniculate nucleus. *Science,* **167,** 1970, 84-86.

Pease, P. L. On color Mach bands. *Vision Research,* **18,** 1978, 751-755.

Pedler, C. Duplicity theory and microstructure of the retina, rods and cones—A fresh approach. In A. V. S. deReuck and J. Knight (Eds.), *Color vision.* Boston, Massachusetts: Little, Brown, 1965. Pp. 52-83.

Pedler, C., & Boyle, M. Multiple oil droplets in the photoreceptors of the pigeon. *Vision Research,* **9,** 1969, 525-528.

Peeples, D., & Teller, D. Color vision and brightness discrimination in two-month-old human infants. *Science,* **189,** 1975, 1102-1103.

Pennal, B. E. Human cerebral asymmetry in color discrimination. *Neuropsychologica,* **15,** 1977, 563-568.

Pokorny, J., Smith, V. C., Verriest, G., & Pinckers, A. J. L. G. *Congenital and acquired color vision defects.* New York: Grune and Stratton, 1979.

Polson, M. C. Spectral sensitivity and color vision in *Tupaia glis.* Doctoral Dissertation. Bloomington, Indiana: Indiana University, 1968.

Polyak, S. *The vertebrate visual system.* Chicago, Illinois: University of Chicago Press, 1957.

Powers, M. K., & Easter S. S.Jr., Wavelength discrimination by the goldfish near absolute visual threshold. *Vision Research,* **18,** 1978, 1149-1154.

Pritz, M. B., Mead, W. R., & Northcutt, R. G. The effects of Wulst ablations on color, brightness and pattern discriminations in pigeons (*Columbia livia*). *Journal of Comparative Neurology,* **140,** 1970, 81-100.

Ratliff, F. *Mach bands: Quantitative studies of neural networks in the retina.* San Francisco, California: Holden-Day, 1965.

Riggs, L. A., & Wooten, B. R. Electrical measures and psychophysical data on human vision. In D. Jameson and L. M. Hurvich (Eds.), *Handbook of sensory physiology,* Vol. VII/4, *Visual psychophysics.* Berlin and New York: Springer-Verlag, 1972. Pp. 690-731.

Ringo, J., Wolbarsht, M. L., Wagner, H. G., Crocker, R., & Amthor, F. Trichromatic vision in the cat. *Science,* **198,** 1977, 753-754.

Rodieck, R. W. *The vertebrate retina*. San Francisco, California: Freeman, 1973.

Romeskie, M., & Yager, D. Psychophysical studies of pigeon color vision. I. Photopic spectral sensitivity. *Vision Research,* **16,** 1976, 501–506.

Romeskie, M., & Yager, D. Psychophysical studies of pigeon color vision. II. The spectral photochromatic interval function in the pigeon. *Vision Research,* **16,** 1976, 507–512.

Rosengren, A. Experiments in colour discrimination in dogs. *Acta Zoologica Fennica,* **121,** 1969, 3–19.

Rumbaugh, D. M. *Language learning by a chimpanzee*. New York: Academic Press, 1977.

Salzle, K. Untersuchungen uber das Farbsehvermogen von Opossum, Waldmausen, Rotelmausen und Eichhornschen. *Zeitschrift fur Saeugetierkunde,* **11,** 1936, 105–148.

Sandell, J. H., Gross, C. G., & Bornstein, M. H. Color categories in macaques. *Journal of Comparative and Physiological Psychology,* **93,** 1979, 626–635.

Saugstad, P., & Saugstad, A. The duplicity theory, an evaluation. *Advances in Ophthalmology,* **9,** 1959, 1–51.

Saunders, R. McD. The spectral responsiveness and the temporal frequency response (TFR) of cat optic tract and lateral geniculate neurons: Sinusoidal stimulation studies. *Vision Research,* **17,** 1977, 285–292.

Schiller, P. H., Stryker, M., Cynader, M., & Berman, N. Response characteristics of single cells in the monkey superior colliculus following ablation or cooling of visual cortex. *Journal of Neurophysiology,* **37,** 1974, 181–194.

Schmidt, I. Some problems related to testing color vision with the Nagel anomaloscope. *Journal of the Optical Society of America,* **45,** 1955, 514–522.

Schneider, B. Multidimensional scaling of color differences in the pigeon. *Perception and Psychophysics,* **12,** 1972, 373–378.

Shefner, J. M., & Levine, M. W. A psychophysical demonstration of goldfish trichromacy. *Vision Research,* **16,** 1976, 671–673.

Sidley, N. A., & Sperling, H. G. Photopic spectral sensitivity in the rhesus monkey. *Journal of the Optical Society of America,* **57,** 1967, 816–818.

Sidman, R. L. Histochemical studies on photoreceptor cells. *Annals of the New York Academy of Science,* **74,** 1958, 812–827.

Silver, P. H. Grey squirrel dichromatic color vision shown by flicker photometry. *Journal of Physiology,* **251,** 1975, 47P.

Sirovich, L., & Abramov, I. Photopigments and pseudo-pigments. *Vision Research,* **17,** 1977, 5–16.

Smythe, R. H. *Animal vision: What animals see*. Springfield, Illinois: Thomas, 1961.

Snodderly, D. M. Visual discriminations encountered in food foraging by a neotropical primate: Implications for the evolution of color vision. In E. H. Burtt Jr. (Ed.), *Behavioral significance of color*. New York: Garland Press, 1979. Pp. 237–279.

Snodderly, D. M., Auran, J., & Delori, F. C. Localization of the macular pigment. *Investigative Ophthalmology and Visual Science Supplement,* **18,** 1979, 80.

Snyder, M., Killackey, H., & Diamond, I. T. Color vision in the tree shrew after removal of the posterior neocortex. *Journal of Neurophysiology,* **32,** 1969, 554–563.

Sokol, S. Cortical and retinal spectral sensitivity of the hooded rat. *Vision Research,* **10,** 1970, 253–262.

Spekreijse, J., Wagner, H. G., & Wolbarsht, M. L. Spectral and spatial coding of ganglion cell responses in goldfish retina. *Journal of Neurophysiology,* **35,** 1972, 73–86.

Stebbins, W. C. *Animal psychophysics*. New York: Appleton, 1970.

Steinberg, R. H. Rod and cone contributions to S-potentials from the cat retina. *Vision Research,* **9,** 1969, 1319–1329.

Steinberg, R. H., Reid, M., & Lacy, P. L. The distribution of rods and cones in the retina of the cat (*Felis domesticus*). *Journal of Comparative Neurology,* **148,** 1973, 229–248.

Stell, W. K., & Lightfoot, D. O. Color-specific interconnections of cones and horizontal cells in the retina of the goldfish. *Journal of Comparative Neurology*, **159**, 1975, 473–502.

Stell, W. K., & Harosi, F. I. Cone structure and visual pigment content in the retina of the goldfish. *Vision Research*, **16**, 1976, 647–657.

Stephenson, P. H. The evolution of color vision in the primates. *Journal of Human Evolution*, **2**, 1973, 379–386.

Stiles, W. S. Increment thresholds and the mechanisms of colour vision. *Documenta Ophthalmologica*, **3**, 1949, 138–165.

Stone, J., & Hoffman, K. P. Conduction velocity as a parameter in the organization of the afferent relay in the cat's lateral geniculate nucleus. *Brain Research*, **32**, 1971, 454–459.

Suthers, R. A. Optomotor responses by echolocating bats. *Science*, **152**, 1966, 1102–1104.

Svaetichin, G. The cone action potential. *Acta Physiologica Scandinavica*, **29**, 1953, 565–600.

Svaetichin, G., & MacNichol Jr., E. F. Retinal mechanisms for chromatic and achromatic vision. *Annals of the New York Academy of Science*, **74**, 1958, 385–504.

Tansley, B. W., & Boynton, R. M. Chromatic border perception: The role of red- and green-sensitive cones. *Vision Research*, **18**, 1978, 683–697.

Thomas, E. Untersuchungen uber den Helligkeits und Farbensinn der Anuren. *Zoologische Jahrbuch Physiologie*, **66**, 1955, 129–178.

Thorell, L. G. The role of color in visual form analysis. Doctoral Dissertation. Berkeley, California: University of California, 1981.

Tigges, J. On color vision in gibbon and orangutan. *Folia Primatologica*, **1**, 1963, 188–198.

Tomita, T. Electrophysiological study of the mechanisms subserving color coding in fish retina. *Cold Spring Harbor Symposium on Quantitative Biology*, **30**, 1965, 559–566.

Tomita, T., Kaneko, A., Murakami, M., & Pautler, E. L. Spectral response curves of single cones in the carp. *Vision Research*, **7**, 1967, 519–531.

Tong, L. Contrast sensitive and color opponent optic tract fibers in the Mexican ground squirrel: Evidence for rod (502 λ_{max}) input. Doctoral Dissertation. Ann Arbor, Michigan: University of Michigan, 1977.

Tribe, D. E., & Gordon, J. G. The importance of colour vision to the grazing sheep. *Journal of Agricultural Science*, **39**, 1950, 313–315.

van der Horst, G. J. C., & Bouman, M. A. Spatiotemporal chromaticity discrimination. *Journal of the Optical Society of America*, **59**, 1969, 1482–1488.

Wagner, G., & Boynton, R. M. Comparison of four methods of heterochromatic photometry. *Journal of the Optical Society of America*, **62**, 1972, 1508–1515.

Wagner, H. G., MacNichol, E. F., Jr., & Wolbarsht, M. L. The response properties of single ganglion cells in the goldfish retina. *Journal of General Physiology*, **43**, 1960, 45–62.

Wald, G., Brown, P. K., & Smith, P. H. Iodopsin. *Journal of General Physiology*, **38**, 1955, 623–681.

Wallman, J. A simple technique using an optomotor response for visual psychophysical measurements in animals. *Vision Research*, **15**, 1975, 3–8.

Wallman, J. The role of retinal oil droplets in the color vision of Japanese quail. In A. M. Granda and J. H. Maxwell (Eds.), *Neural mechanisms of behavior in the pigeon*. New York: Plenum Press, 1979. Pp. 327–351.

Walls, G. L. The visual cells of the white rat. *Journal of Comparative Psychology*, **18**, 1934, 363–366.

Walls, G. L. *The vertebrate eye and its adaptive radiation*. Bloomfield Hills, Michigan: The Cranbrook Institute of Science, 1942.

Walls, G. L., & Heath, G. G. Neutral points in 138 protanopes and deuteranopes. *Journal of the Optical Society of America*, **46**, 1956. 640–649.

Walton, W. E., & Bornemeier, R. W. Color discrimination in rats. *Journal of Comparative Psychology*, **28**, 1938, 417–436.

Wartzok, D., & McCormick, M. Color discrimination by a Bering sea spotted seal, *Phoca largha*. *Vision Research*, **18**, 1978, 781–784.

Watson, J. B., & Watson, M. I. A study of the responses of rodents to monochromatic light. *Journal of Animal Behavior*, **3**, 1913, 1.

Weale, R. A. Bleaching experiments of eyes of living grey squirrels (*Sciurus carolinensis leucotis*). *Journal of Physiology*, **127**, 1955, 585–591.

Weiskrantz, L. Contour discrimination in a young monkey with striate cortex ablation. *Neuropsychologica*, **1**, 1963, 145–164.

West, R. W., & Dowling, J. E. Anatomical evidence for cone and rod-like receptors in the gray squirrel, ground squirrel, and prairie dog. *Journal of Comparative Neurology*, **159**, 1975, 439–460.

Wiesel, T. N., & Hubel, D. H. Spatial and chromatic interactions in the lateral geniculate body of the rhesus monkey. *Journal of Neurophysiology*, **29**, 1966, 1115–1156.

Wojtusiak, R. J. Uber den Farbensinn der Schildkroten. *Zeitschrift Vergleichende Physiologie*, **18**, 1933, 393–436.

Wolbarsht, M. The function of intraocular filters. *Federation Proceedings*, **35**, 1976, 44–50.

Wolken, J. J. *Photoprocesses, photoreceptors and evolution*. New York: Academic Press, 1975.

Wright, A. A. The influence of ultraviolet radiation on the pigeon's color discrimination. *Journal of the Experimental Analysis of Behavior*, **24**, 1972, 325–337. (a)

Wright, A. A. Psychometric and psychophysical discrimination functions for the pigeon. *Vision Research*, **12**, 1972, 1447–1464. (b)

Wright, A. A. Bezold-Brücke hue shift functions for the pigeon. *Vision Research*, **16**, 1976, 765–774.

Wright, A. A. Construction of equal-hue discriminability scales for the pigeon. *Journal of the Experimental Analysis of Behavior*, **29**, 1978, 261–266.

Wright, A. A., & Cummings, W. W. Color-naming functions for the pigeon. *Journal of the Experimental Analysis of Behavior*, **15**, 1971, 7–17.

Wright, W. D. *Researches on normal and defective colour vision*. St. Louis, Missouri: Mosby, 1947.

Wyzecki, G., & Stiles, W. S. *Color science*. New York: Wiley, 1967.

Yager, D. Behavioral measures and theoretical analysis of spectral sensitivity and spectral saturation in the goldfish *Carassius auratus*. *Vision Research*, **7**, 1967, 707–727.

Yager, D. Effects of chromatic adaptation on saturation discrimination in the goldfish. *Vision Research*, **14**, 1974, 1089–1094.

Yager, D., & Jameson, D. On criteria for assessing type of colour vision in animals. *Animal Behaviour*, **16**, 1968, 29–31.

Yager, D., & Sharma, S. C. Evidence for visual function mediated by anomalous projection in goldfish. *Nature (London)*, **256**, 1975, 490–491.

Yarczower, M., & Bitterman, M. Stimulus-generalization in the goldfish. In D. Mostofsky (Ed.), *Stimulus generalization*. Stanford, California: Stanford University Press, 1965. Pp. 179–192.

Yates, T. Chromatic information processing in the foveal projection (*Area striata*) of unanesthetized primate. *Vision Research*, **14**, 1974, 163–179.

Yazulla, S., & Granda, A. M. Opponent-color units in the thalamus of the pigeon (*Columba livia*). *Vision Research*, **13**, 1973, 1555–1563.

Yolton, R. L. The visual system of the western gray squirrel: Anatomical, electroretinographic and behavioral studies. Doctoral Dissertation. Austin, Texas: University of Texas, 1975.

Yolton, R. L., Yolton, D. P., Renz, J., & Jacobs, G. H. Preretinal absorbance in sciurid eyes. *Journal of Mammalogy*, **55**, 1974, 14–20.

Young, R. W. The organization of vertebrate photoreceptor cells. In B. R. Straatsma, M. O. Hall, R. A. Allen, & F. Crescitelli (Eds.), *The retina: Morphology, function and clinical characteristics.* Berkeley, California: University of California Press, 1969, Pp. 177–210.

Young, R. W. Visual cells and the concept of renewal. *Investigative Ophthalmology,* **15,** 1976, 700–725.

Zeki, S. M. Colour coding in rhesus monkey prestriate cortex. *Brain Research,* **53,** 1973, 422–427.

Zeki, S. M. Colour coding in the superior temporal sulcus of rhesus monkey visual cortex. *Proceedings of the Royal Society of London,* **197,** 1977, 195–223.

Zeki, S. M. Uniformity and diversity of structure and function in rhesus monkey prestriate visual cortex. *Journal of Physiology,* **277,** 1978, 273–290.

Zeki, S. M. The representation of colours in the cerebral cortex. *Nature (London),* **284,** 1980, 412–418.

Zrenner, E., & Gouras, P. Blue-sensitive cones of the cat produce a rodlike electroretinogram. *Investigative Ophthalmology and Visual Science,* **18,** 1979, 1076–1081.

Zwick, H., & Robbins, D. O. Is the rhesus protanomalous? *Modern Problems in Ophthalmology,* **19,** 1978, 238–242.

Author Index

Numbers in italic indicate the page on which the complete reference is listed.

Subject Index